Michéle Tancman Candido da Silva

The impact of the Digital Cities Program and other RJ initiatives

Michéle Tancman Candido da Silva

The impact of the Digital Cities Program and other RJ initiatives

Alternatives to enhance the territory in order to complement the organization of real cities

ScienciaScripts

Imprint

Cover image: www.ingimage.com

This book is a translation from the original published under ISBN 978-3-330-75995-4.

Publisher:
Sciencia Scripts
is a trademark of
Dodo Books Indian Ocean Ltd. and OmniScriptum S.R.L publishing group

120 High Road, East Finchley, London, N2 9ED, United Kingdom
Str. Armeneasca 28/1, office 1, Chisinau MD-2012, Republic of Moldova, Europe
Printed at: see last page
ISBN: 978-620-8-31461-3

SUMMARY

PRESENTATION

Digital cities are emerging as an alternative way of enhancing the territory in a way that complements the organization of real cities. In Brazil, the Ministry of Communication's Digital Cities Project seeks to create the means for citizens to access, use, produce and disseminate information and knowledge in order to participate effectively and critically in the information society.

The main objective of this project is to analyze the social, economic, political and cultural impacts of the creation of digital cities in the territory of the cities in the state of Rio de Janeiro covered by the Digital Cities Pilot Program of the Ministry of Communication and the public policy PAC 2 - Rio de Janeiro 2013 and to verify other initiatives already implemented by other policies, in the context of the systems of objects and practices that represent the process of digital inclusion as one of the prerequisites for exercising citizenship in the midst of socio-spatial contradictions.

Specific objectives: **to** identify the impact of levels of control on the dynamics of real urban spaces by evaluating the strategies used to disseminate digital city networks; to identify the effects of digital inclusion and exclusion, especially with regard to "democracy" of access to new communication networks. This issue is important because it explains social exclusion, Identifying and mapping the potential of cities themselves and their capacity to implement local development with a technical and political vocation.

The starting point for the proposed research methodology is a survey of digital initiatives in the cities of Engenheiro Paulo de Frontin, Maricà and Sâo José do Vale do Rio Preto in the state of Rio de Janeiro, which were included in the Pilot Project - 2012, and the cities included in PAC2 - 2013 - Bom Jardim, Casimiro de Abreu, Itaocara, Miguel Pereira, Paraiba Sul and Sâo Francisco de Itabapoana, Casimiro de Abreu, Itaocara, Miguel Pereira, Paraiba do Sul, Paraty, Pirai, Santo Antônio de Pàdua, Sâo Francisco de Itabapoana, Sâo Joâo da Barra, Silva Jardim and Vassouras, all linked to the Public Policy of the Ministry of Communication's Digital Cities Program.

The theoretical methodology will be based on the concepts of territory, networks and the debate around the notion of cyberspace and digital cities. Secondary data from government agencies such as IBGE, MCT, ANATEL and others will be analyzed in order to subsidize the program's research.

Systematic and participatory observation was carried out through interviews, mainly with the managers of the Digital Cities covered by the program.

This study is justified in order to understand the implications and impacts that could occur in existing socio-spatial conflicts. The aim is to compare these results with information on the administrative and operational structure of the public authorities responsible for territorial management, thus shaping the general lines of understanding and implementing the digital city.

Will these programs respond to the needs of citizens and local communities? What strategies have been adopted to achieve social, political and economic development?

PART 1

DEVELOPMENT

THEORETICAL-METHODOLOGICAL RESEARCH

1. Technological advances favor Digital Cities initiatives

Technological advances in the areas of information technology and telecommunications are one of the supports for Digital Cities initiatives. In addition to the development of digital content, forms of social relations are emerging, as well as a field for the development of sociability and political and cultural activism, which are expressed through objects with a strong technical-scientific-informational content.

Therefore, the impact of these cities on the territory is being appropriated by the logic of capital on different scales and also by public policy initiatives, whether municipal, regional or national, such as different forms of digital inclusion, construction of telecentres, modernization of public services, construction of fibre optic and hybrid networks, Internet access and others.

This causes changes, with the effect of enlarged or excluded spaces, based on a globalized process of political, social and economic relations that are established, establishing a constant local territorial (re)ordering.

Cyberspace tends to represent the space of information flow, constructed by technology, taking on an interactive or control configuration, that is, a rigid control of the social fabric which, when it takes on a material form, according to Alves (2003), corresponds to the functions of social control and hierarchical power.

In this case, according to the analysis made by Castells (1999), communication networks would have a fertile field in the dissemination of culture, discourse and the construction of a collective identity, mainly from the point of view of identity and the relations between state - nation and democracy, based on the study of the relations between cultural identity, social and political movements.

In the space where communication networks flow, it is possible to process and store information thanks to technology. In this way, network users benefit from communicating with each other, creating "networks", searching for data and empowering social movements. There is also the availability to trigger globally interconnected databases, government and private interests in social networks.

The space of flows of image, sound, information and sociality defined by cyberspace expresses a "material organization of social practices of shared time that works through flows" (CASTELLS, 1999, p.436). However, it is necessary to avoid the counter-effects of creating these new spaces for more complex analysis, because instead of the progress that can be made in a given space and its surroundings, there could be situations of social exclusion. At the forefront is the fight against exclusion, not only digital exclusion but also exclusion from knowledge production networks, which are portals for access to content and opportunities. In this way, it is understood that cyberspace exists but that access to it is not for everyone. In Brazil, according to data from the IBGE (ICT Households 2014), internet access in households reached 85.6 million Brazilians, equivalent to 49.4% of the population.

Assuming that cyberspace allows for the expansion of strategies to adapt socio-spatial accumulation patterns, those outside it are doomed to the existing mismatch in their political order. They are almost always hostage to the principles of sovereignty and the new economic and social formations represented by various hegemonic groups.

On the other hand, this innovation in the construction of Digital Cities and access to the web provides opportunities for "those techniques" cited by Milton Santos (2014) as "sweets" to contribute more effectively to the service of man.

Milton Santos (2014)i states the following:

> However, the same materiality that is currently being used to build a confusing and perverse world could become a condition for building a more humane world. All that is needed is to complete the two great changes currently taking place: the technological change and the philosophical change in the human species. ··The great technological change is the emergence of information technology, which, unlike machine technology, is constitutionally divisible, flexible and docile, adaptable to all media and cultures, even if its current perverse use is subordinated to the interests of big capital. But when their use is democratized, these sweet techniques will be at the service of man (SANTOS, 2014).

The definition of networks coined by Milton Santos, as one of his conceptual matrices, takes into account the social and political nature of technical information flows. The emergence of digital city networks responds to material and immaterial stimuli that make them eminently geographic. As such, the analysis of Digital Cities involves the dynamics of networks.

The transformative actions needed to integrate and achieve social synergies involve the creation of communities of collaborators and the establishment of partnerships, the dynamics of which are complex and require management planning. In this case, the implementation of national programmes involves the hierarchical control of digital city networks.

From this perspective of analysis, it can be argued that real cities acquire a new content from cyberspace, translated as digital cities.

We understand that the digital city incorporates much more than a "platform" and users. It is one of the concrete manifestations of the information network society.

Another aspect to be analyzed is the digital city as a tendency towards regional and local aspects, a counterpoint to the globalizing tendencies characteristic of the diffusion of the Internet and other information and communication technologies. Some authors, such as Jordi Borja (1997), talk about a new "social platform" that aims for something in-between, a kind of "inter-community network" that would be on the way between the Global and the Local, based on coordination between movements of community organization of networks through a communication system that ensures connectivity and global flows of people, goods, information and, above all, human resources capable of producing and managing a new economic system.

The current technological transformations are creating an era of communication and social networks in which digital cities are emerging as an alternative way of enhancing the territory, introducing new scales of social relations. It is a fact that the territory of Digital Cities depends on local infrastructure (cables, fiber optics, satellites, telephone systems, antennas, software and hardware, etc.) to integrate material and virtual space.

As a result, cities that lack resources and an urban planning policy aimed at implementing the basic requirements for a digital city will be prevented from participating in access to information networks, which could lead to growing technological inequality.

For Kurbalija and Gelbstein (2005), according to figure 1, the Internet infrastructure, the basis for the realization of projects, is divided into at least three groups (telecommunications infrastructure, technical standards and content and application standards) whose basis is based on the techniques used in the space, represented by figure no. 6, below.

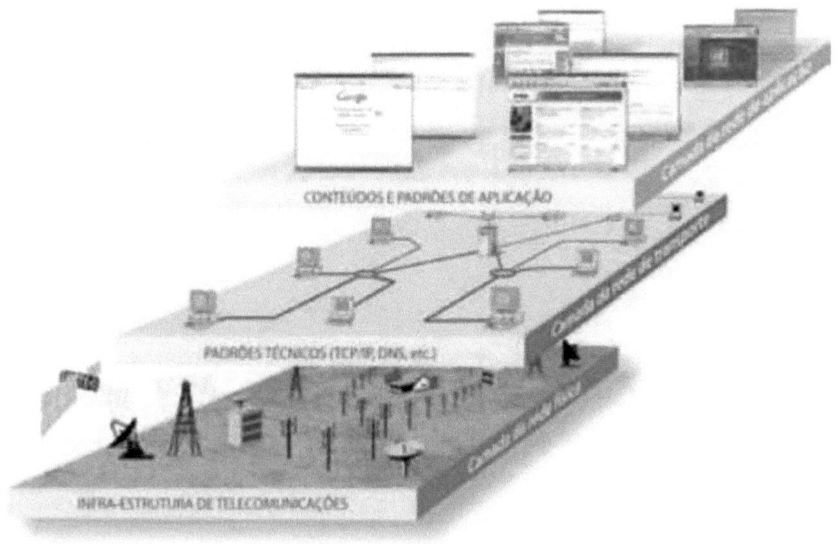

Figure 1 - telecommunications infrastructure, technical standards and content and application standards Source: Kurbalija, J. and Gelbstein, E. Internet governance, issues, actors and decisions, 2005 p.37.

Regardless of the stage the digital city is at, it reflects the dynamism of the real city and presupposes a complex, interactive, unstable and self-organizing order that relies on the disorder expressed in access to the network and the various socialities that are present there. It is a construction process that needs an infrastructure, as this is certainly a limiting factor for its implementation. A network of services and goods available on the Internet that allows society to have direct contact with the virtual space depends on the strategy of supporting investments that make it easier for the less well-off to enter the new technologies.

Digital Cities have an effect on the real space of the city, as the restrictions imposed by distance and time can be overcome. It's like a compression of space and time, as Harvey (1993) points out. Therefore, the higher the transmission speed, for applications that require the transfer of large amounts of data, the more agile social interactions will be.

Digital cities are strategically important in cyberspace. As well as concentrating a large number of services and attractions for people to frequent them, they have created mechanisms to universalize access to digital services on different scales, from the construction of telecentres, such as those in several cities, to programs aimed at investing in digital literacy and inclusion.

Considering the social and economic inequalities of the real city, where most people have difficulty even accessing public spaces, it is understood that digital inclusion is a goal, a reality to be achieved and it is not an easy task, because those who already have access to the net do not necessarily benefit from broadband and constantly developing digital content.

The Digital City is a reflection or the very social, economic and political development of the Royal City.

Pires (2004) explains that the digital city is an acceleration and represents the expression of the territorial restructuring of the new regime of capital accumulation under the sign of the capitalist network society. It is the digital expression of the information age. This era has been increasingly influenced by the spatial expansion of the territorial structures of accumulation of the capitalist network society.

The definition given by Pires reinforces the understanding that the digital city needs an infrastructure capable of articulating the dispersed flows mediated by technology

Despite the optimism about the implementation of digital city networks, one cannot fail to mention that the technologies used, as an important instrument, if not fundamental to the formation of digital cities, at the same time as enabling the spread of the most sophisticated technological resources to a certain social segment, exclude the majority of the population from accessing them.

In general, the projects seek to articulate the participation of society and open up a reflection on the influence of exchanges of experience capable of impacting, positively or negatively, on the territory of real cities.

The emergence of the construction of digital cities is based on the contradictions that exist in geographical space. Therefore, the fundamental preconditions for access to information, knowledge, transparency, reducing bureaucracy and all the other forms of development announced by these programs, in order to create a fairer society, remain the same, with or without new technologies. In this way, a policy is also needed to combat hunger, poverty, violence, illiteracy, homelessness, among many other important and fundamental needs to improve the quality of life, i.e. to offer "new" opportunities for development, transforming representative democracies into participatory democracies.

Based on this scenario, the creation of digital cities or digital city networks, according to Santos (1996, p215), are the vehicles of a dialectical movement which, on the one hand, opposes territory and place to the world; and, on the other hand, confronts place with territory taken as a whole", since their "existence is inseparable from the question of power". Some impacts on the territory can be seen from the implementation of ICTs (Information and Communication Technologies) and the creation of digital cities from the increase in information such as:

- The allocation of public resources aimed at creating an information network and dissemination mechanisms that prioritize the joint work of the public and private sectors involved in various issues, such as the environment.
- The expansion of citizenship through greater participation by the various communities in the definition of public policies, respecting, perhaps, cultural diversities and educational differences.
- The spread of a virtual information network with no defined territorial base, made up of various social groups with addresses linked to the World Wide Web (the Internet).
- The resizing of actions and decision-making processes that will allow decentralization and the formation of a new "urban hierarchy" based on new organizational processes for the exchange of information and technical knowledge between places.
- The construction of platforms for the development of public policies and technological infrastructures, based on free software, whose founding premise of the movement inspired by the concepts of freedom of the "Free Software Foundation" is to deepen the construction of a concrete alternative that seeks to insert the technological issue in the context of a world with social inclusion and equal access to technological advances.
- Controlling urban mobility

In this sense, digital cities are emerging as an alternative way of enhancing the territory in a way that complements the organization of "real cities". The digital city is the basis for building a smart city, i.e.

a city that is both creative and sustainable. Technology is the basis for its planning process with the participation of citizens. These are cities whose concept is based on physical apparatus, cabling, for example. In other words, they are the infrastructural apparatuses that are installed in a city or part of a city. These cities enable a series of activities with predefined programs in a semi-automatic way and are designed to meet the needs of their inhabitants, since all the infrastructure that makes them up is aimed at reaching every home, business and workplace. A substantial difference is that when a smart city is identified, the infrastructure aspects resulting from a series of traditional combinations of providers, networks and equipment in its physical space are considered so that all residents can enjoy certain services offered remotely.

A European conception of the smart grid (2014) refers to a set of urban planning and technological solutions aimed at sustainable development and quality of life. It is the application of information technology (IT) to the electricity system, for example, integrated with communication systems and automated network infrastructure in order to make the process more efficient.

On a global scale, some initiatives are benchmarks and have been designed to serve the networked community, virtually created according to the needs of each community.

2. Initiatives to consolidate Digital Cities in Rio de Janeiro

2. 1 - The genesis of the implementation of the Digital Cities Project promoted by the Ministry of Communications.

The Digital Cities Project is one of the public policy initiatives in Brazil promoted by the Ministry of Communications, initiated by Ordinance No. 376 of August 19, 2011, published in the Official Gazette of the Union on August 22, 2011, which aims to establish the Digital Cities Implementation and Maintenance Project. This document defines the concept of Digital Cities as local digital communication networks in Brazilian municipalities aimed at digital inclusion and has the following objectives.

- Improving quality and transparency in public management;
- The democratization of access;
- Fostering a creative and sustainable economy;
- Content creation and development;
- Building collaborative environments in open networks.

Based on the National Broadband Program (PNBL[2]), a public policy instituted by Decree No. 7.175 of May 12, 2010, created by the Federal Government with the aim of expanding Internet access in the country, the MC - Ministry of Communications prepared a Public Call Notice No. 1/2012 for the centralized execution of the Digital Cities Pilot Project.

One of the great possible advantages advertised is to offer a donation conditional on the transfer of ownership of the metropolitan network installed by the donor in the municipality.

The metropolitan network would be based on optical fiber, composed of hardware, software and complementary accessories, allowing government agencies, public institutions and spaces providing public services to the population to be interconnected and have broadband access to the Internet, in addition to providing free access to citizens in specific locations. (Public Call No. 1/2012).

This Metropolitan network, considered an infovia, would include the supply and installation of equipment, services and software necessary for its operation, as well as technical support and training of local staff to manage the infrastructure.

The Ministry of Communications, through TV MiniCom (2011) and according to Qrcode No. 01,

2 The intention of the PNBL was to massify access to broadband Internet connection services; accelerate economic and social development; promote digital inclusion; reduce social and regional inequalities; promote the generation of employment and income; expand e-Government services, facilitating the use of state services; promote the training of the population in the use of Information and Communication Technologies (ICT). However, according to an analysis carried out by a specialist and based on an evaluation report of the PNBL carried out by the Federal Senate's Committee on Science, Technology, Innovation, Communication and Informatics, it can be seen that the National Broadband Program did not meet its targets for 2014.

Digital Cities Project, summarizes the project as one of the priorities of the Ministry of Communications.

Digital Cities Project (Published on May 20, 2011)

Source: TV MiniCom (2016)

2.2 The methodology for implementing the Digital Cities Pilot Project

The methodology used to develop the Pilot Project, at first and then in the other cities, considered six phases until its conclusion. All the phases require partnerships between the Union and the municipalities, which are shown in Table 01. - Phases for completing the Digital Cities Project.

Phases	Actions
1ª Phase	Through a public call for proposals. The choice followed criteria that favored municipalities with low broadband connection density, lower development indices, small populations, in the North and Northeast regions and that formed consortia. The possibility of providing the connection, preferably by Telebràs, was also required;
2a Phase	Tendering and contracting of the company(ies) that will be responsible for deploying the optical network, including the supply and installation of equipment and software
	necessary for its operation, technical support, technology transfer and assisted operation for a period of six months, with the aim of the municipality taking ownership of the technology implemented.
3ª Phase	Formalization of cooperation agreements involving the Union and the beneficiary municipalities, through which joint efforts will be established, with a distribution of responsibilities, to make the project viable. In these cooperation agreements, responsibilities will be distributed as follows: • **Uniao** - undertakes to provide for the installation of the fiber optic ring - metro ethernet network, the connection equipment and to donate it free of charge. It will also provide technology transfer and assisted operation of the network. • **Municipalities -** undertake to provide a local management team to be trained to be able to monitor the project; to contribute to the logistics for setting up the connection infrastructure; to share responsibility for implementing actions related to digital inclusion;

	and to provide the necessary information for installing the infrastructure.
4a Phase	Formalization of a term of donation with charges for the network infrastructure and connection equipment to the municipalities, in which they commit themselves to the maintenance and upkeep of the network and equipment for a period of 3 years, guaranteeing their use in public spaces, and to adherence to digital inclusion actions that include e-government, training of servers and monitors and provision of appropriate content and applications.
4a Phase	Implementation of e-government applications in the areas of financial management, taxation, health and education.
6a phase	Training of municipal civil servants in the management and use of the installed network, use and management of e-government applications.

Table 01. - Phases for completing the Digital Cities Project

Source: Digital Cities documents (2016)

According to the MC (2016), the pilot project for a Digital City makes it possible to modernize city management by setting up network connection infrastructure between public bodies, public management applications, training civil servants and providing free public internet access for the population. It provides the community with access to government services, as well as the digital inclusion of Brazilian municipalities, resulting in local development.

This project has partnerships with the Ministry of Planning, Telebrâs, Inmetro and the BNDES. We can really conclude that this project would effectively contribute to the development of the geographical area where it is implemented.

Through Electronic Tender Notice No. 12/2012-MC, integration companies were contracted to be responsible for the physical installation of equipment, network cables and the entire infrastructure in each municipality.

In the first instance, after selecting the integrating companies responsible for the project, a contract is signed including the preparation of an executive project, which includes the entire infrastructure network of works, services, equipment and installations.

Secondly, the inspection and approval phases begin.

All the executive projects should be accepted and approved by the Ministry of Communications and the concession company, and only then will Inmetro begin the inspection process to check that the installations comply with its technical standards.

It is then the responsibility of the municipality participating in the program to receive the license called

Private Limited Service - SLP .[3]

Once the grant has been received, the town hall will contract a server *link* for all internet access.

The contracted company will be responsible for maintenance and support for at least the first six (6) months. After this period, the municipality must take over the management and operation of the infovia. Even so, the municipalities will have a guarantee of 3 (three) years from the start of assisted operation of the equipment and software, as explained by Cidades Digitais (2016).

In this modality, the Digital Cities Project consists of: Implementing a metropolitan network, in the urban core, based on optical fiber, composed of *hardware*, *software* and complementary accessories for the installation of the network, including an infrastructure management solution, with assisted operation for six (6) months, and the subsequent donation with charges of the implanted infrastructure; installation of e-government applications, with support for migration, training, hosting and assisted operation; and training for public servants on how the network works and how to use Information Technology (ICT) tools to improve public management. (CIDADES.DIGITAIS, 2016).

2.2.1 The municipalities included in the Rio de Janeiro State Digital Cities Pilot Project and the selection criteria.

The 80 municipalities selected for the pilot project are from different Brazilian states. However, according to Castro (2015) it appears that of the 80 municipalities, 3 have already been de-registered, 50 have the network lit up, and 27 may have the network completed.

The evaluation committee explained that managerial and technical capacity, the sustainability of the project and the expansion of the network contributed to the choice of cities. To give you an idea, more than 50% (42) of the municipalities tendered for the connection.

The cities were classified based on the criteria in the public notice (2012) and judged according to Table 2 - Scoring for the Classification of Proposals.

Scoring Table for the Classification of Proposals Criteria	**I Scoring**
A - Project Presentation	
Problem description	Up to 3 points.
Relevance to the municipality or administrative region of the Federal District	Up to 3 points.
Implementation of e-government programs and their relationship with other actions and projects related to municipal management	Up to 10 points.
Clarity and relevance of the objectives, goals and expected results and	Up to 6 points.

3 Resolution No. 617, of June 19, 2013, states: SLP is a telecommunications service, of restricted interest, exploited nationally and internationally, under the private regime, intended for the use of the performer itself or provided to certain groups of users, selected by the provider according to criteria established by it, and which covers multiple applications, including data communication, video and audio signals, voice and text, as well as the capture and transmission of scientific data related to Earth exploration by satellite, aid to meteorology, meteorology by satellite, space operations and space research.

relevance of the indicators, describing the expected economic and social impact following the implementation of the Digital City.	
Municipality's IFDM Expression for scoring: defined by the formula: 10*(1-IFDM)	Up to 7 points.
Household Density of Broadband Access in the Territory of the Municipality or Administrative Region of the Federal District. Expression for scoring: defined by the formula: 10 *(1-DBL)	Up to 10 points.
Project implemented through a consortium of municipalities.	5 points
Per capita current income of the municipality or administrative region of the Federal District of less than R$1,200.00. Expression for scoring: defined by the formula: (1.200,00 - rec.cor.p.cap)/300. An additional point if the municipality belongs to the g100 defined by the National Front of Mayors (FNP).	Up to 4 points
Municipality in the North or Northeast region	10 Points
Population of the municipality up to 50 thousand inhabitants	10 Points
Subtotal A:	Up to 68 points.
B - Managerial and Technical Capacity of the Tenderer	
Applications already in use for public management and digital inclusion	Up to 5 points.
Permanent staff available for training and who will be involved in project management	Up to 10 points.
Identification of local infrastructure for project implementation	Up to 10 points
Subtotal B:	Up to 25 points.
C. Project sustainability and expansion	
Proposal for the project's own funding mechanisms	Up to 10 points
Possibility of establishing partnerships to maintain/operate the project	Up to 20 points
Municipality or Federal District has signed a commitment with a private or public operator to provide *backhaul* at monthly prices equal to or less than (not cumulative):	> R$ 230,00 - 0 points R$ 230,00 - 5 points R$ 200,00 - 6 points R$ 150,00 - 7 points
Subtotal C:	Up to 37 points.
Total:	Up to 130 points.

Table 2 - Scoring for the Classification of Proposals.

Source: MC 01/2012 - Digital Cities Project (2016)

According to the criteria adopted, municipalities in the North and Northeast would already be guaranteed 10 points and the creation of a municipal consortium would increase them by 5 points.

Another point that stood out was in relation to the municipalities that had a low density of broadband connections, as well as the possibility of providing a connection, preferably by Telebrâs.

In the state of Rio de Janeiro, only three municipalities were included in the Pilot Project: Engenheiro Paulo de Frontin, Maricà and Sâo José do Vale do Rio Preto, as shown in Figure 2 - Digital Cities selected for the Pilot Project.

Figure 2 - Digital Cities selected for the Pilot Project

Source: Digital Cities documents (2016)

It is important to note that in item "A" - presentation of the project in the Scoring Table for the Classification of Proposals - municipalities with up to 50,000 inhabitants would already be guaranteed 10 points out of the 68 points for the classification of the proposal. Maricà did not score in this item, since its population, according to the 2010 census, was 127,461 inhabitants, unlike Engenheiro Paulo de Frontin and Sâo José do Vale do Rio Preto.

The estimated population in 2015 can be seen in Table 3, entitled Estimated population of each municipality in the state of Rio de Janeiro.

Municipality	Estimated population (2015)

Engenheiro Paulo de Frontin	**13.626**
Maricà	**146.549**
Sâo José do Vale do Rio Preto	**20.916**

Table 3: Estimated population of each municipality in the state of Rio de Janeiro. Source: IBGE (2015)

So it can be concluded that Maricà benefited from other criteria in the selection process.

2.2.2 The current status of Rio de Janeiro's Digital Cities Pilot Projects

The construction of infrastructure for the use of technologies in these cities would be an opportunity to disseminate important projects, especially to establish significant connections with other municipalities or even expand the technological base, but the current status of these pilot projects in Rio de Janeiro (2016) is still embryonic. Table 4 shows that in Maricà and Sâo José do Vale do Rio Preto, the infrastructure has not yet been started (2016) and Engenheiro Paulo de Frontin is still waiting for the project to begin.

DIGITAL CITIES - RIO DE JANEIRO		
Selection	**Municipality**	**Status**
Pilot Project	Engenheiro Paulo de Frontin	Awaiting start of implementation
	Marica	Infrastructure not started.
	Sao José do Vale do Rio Preto	Infrastructure not started.

Table 4 - Pilot Project

Source: (CIDADES.DIGITAIS@COMUNICACOES.GOV.BR, 2016),

The Pilot Projects are expected to be completed by the end of 2015. Castro (2015) explains that of the three participating municipalities in Rio de Janeiro, two have already contracted the connection link and Engenheiro Paulo de Frontin is in the process of being inspected by Inmetro, which means they are checking that the installations comply with their technical standards.

The installation of access points and fiber optics has not begun in any municipality in the state of Rio de Janeiro. Only one municipality has been granted an SLP - Private Limited Service license

In 2013, the Digital Cities Project gained momentum as part of the Federal Government's Growth Acceleration Program (PAC). As part of PAC2, the project's implementation and results are also monitored by the *Ministry of Planning, Budget and Management* (MPOG).

One of the criteria adopted for the selection of cities is that they must have up to 50,000 inhabitants. Another criterion concerns its logistics, as the city should preferably be no more than 50 km from the Telebrás backbone (the Ministry of Communications will carry out this check, based on the planning for the deployment of Telebrás networks) or have a signed commitment, which can be verified, with a private operator to provide an internet connection;

2.3 - Municipalities in the State of Rio de Janeiro included in the Digital Cities project by PAC II

PAC2[4] is a public policy of the Ministry of Planning, Management and Budget (MPOG) that incorporates more social and urban infrastructure actions in relation to PAC1 and, among other priorities, investment in the construction of public facilities that provide the population with comfort, safety and access to essential services such as nurseries, basic health units, spaces for sport, culture and leisure. (MPOG, 2016)

With this in mind, the MPOG's Digital Cities Program launched a new call for proposals with the aim of selecting other cities. Several Brazilian municipalities[5] sent in proposals to take part in the second phase of the Digital Cities Program, a total of 1901 (MC, 2015). Of these 1901, only 260 were selected. In the state of Rio de Janeiro, there are a total of 12 cities, namely: Bom Jardim, Casimiro de Abreu, Itaocara, Miguel Pereira, Paraiba do Sul, Paraty, Pirai, Santo Antônio de Pàdua, Sâo Francisco de Itabapoana, Sâo Joâo da Barra, Silva Jardim and Vassouras. They are shown spatially in Figure 3. - List of municipalities selected by PAC2.

4 PAC2 is a public policy that was planned for the period from 2011 to 2014.
5The full list of cities in Brazil is in the document Final result - selected municipalities . Available at < http://www.comunicacoes.gov.br/formularios-e- requerimentos/cat view/22-acoes/27-cidades-digitais >. Accessed Jan. 2016

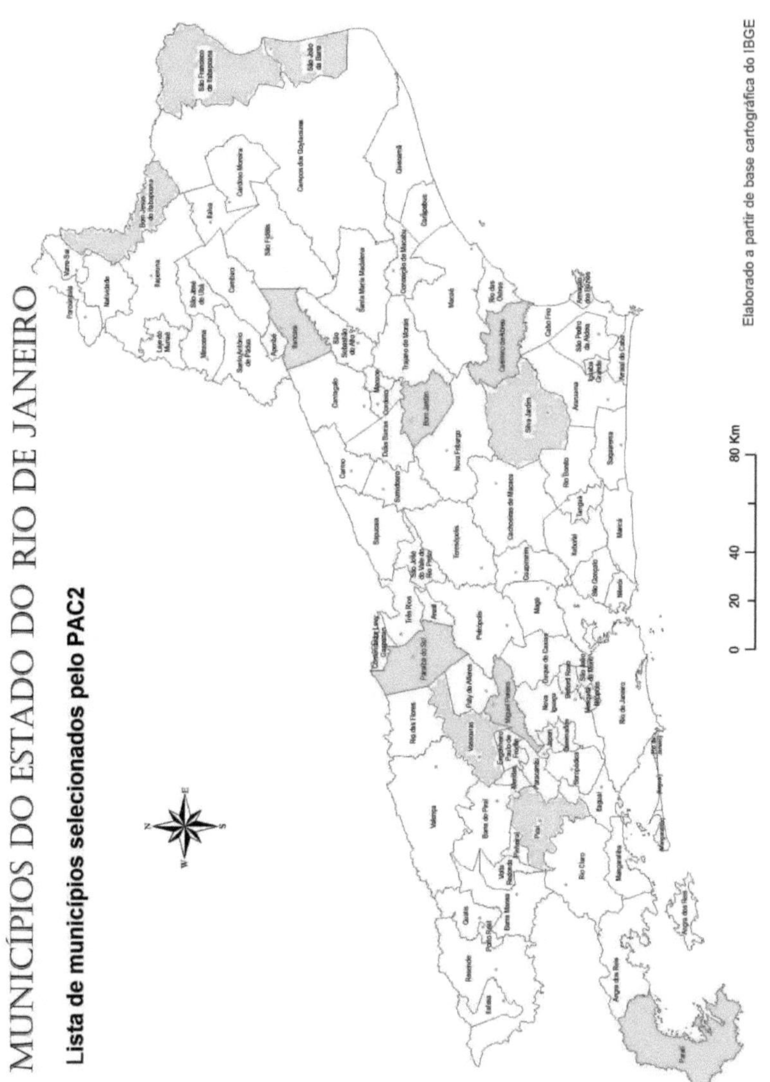

Figure 3 List of municipalities selected by PAC2 Source: Digital Cities documents (2016)

All the new municipalities selected have fewer than 50,000 inhabitants, according to Table 5 - Estimates of the resident population in Brazilian municipalities.

Municipality	Estimated population (2015)
Bom Jardim	**26.278**

Casimiro de Abreu	**40.305**
Itaocara	**22.779**
Miguel Pereira	**24.842**
Paraiba do Sul	**42.356**
Paraty	**40.478**
Pirai	**27.838**
St. Anthony of Pàdua	**41.178**
Sâo Francisco de Itabapoana	**41.291**
Sâo Joâo da Barra	**34.583**
Silva Jardim	**21.307**
Brooms	**35.432**

Table 5 - Estimated resident population in Brazilian municipalities with a reference date of July 1, 2015.

Source: IBGE (2016)

In this new phase of the Digital Cities Program, the investments planned by the cities selected by PAC2 in Rio de Janeiro were released by the MPOG according to Table 6. The planned investments.

Municipality(ies)	Planned investment
Bom Jardim	R$770.980,33
Casimiro de Abreu	R$842.187,31
Itaocara	R$709.169,31
Miguel Pereira	R$1.387.111,83
Paraiba do Sul	R$1.267.858,79
Paraty	R$1.050.686,93
Pirai	R$907.753,83
St. Anthony of Pàdua	R$1.338.880,40
Sâo Francisco de Itabapoana	R$610.044,96
Sâo Joâo da Barra	R$547.505,97
Silva Jardim	Not informed
Brooms	Not informed

Table 6 - Planned investments approved

Source: PAC Infrastructure (2016) - Reference Date 06/30/2015

It is important to note that Silva Jardim and Vassouras, despite belonging to the

The selected cities do not provide any information on the planned investment.

Based on the budget forecast for each municipality, it was hoped that the projects would at least be in development, but in fact what we see (2016) in the state of Rio de Janeiro is that there has been no progress, as shown in Table 7 - Status of the Digital Cities Projects in Rio de Janeiro covered by PAC2.

DIGITAL cities - RIO DE JANEIRO		
Seven	Municipality	Status
PAC2	Bonn Garden	Executive Project OS not issued.
	Casimiro de Ahreu	Executive Project OS not issued.
	ita oc ara	Executive Project OS not issued (CADiNI debit)
	Miguei Pereira	Executive project awaiting analysis
	Paraiba do Sul	Executive Project not received
	Paraty	Executive project awaiting analysis
	Pirai	05 of the Executive Project not issued [CADINI indebtedness)
	Saint Anthony of Padua	Executive Project not received
	Sao Francisco de Stahapoana	Executive project not received
	SaoJoao da Barra	Executive project not received
	Siiva Jardim	OS of the Executive Project not yet issued (debit) CADINi)
	Brooms	Executive project not received

Table 7 - Status of projects in the Digital Cities in Rio de Janeiro covered by PAC2.

Source: (CIDADES.DIGITAIS@COMUNICACOES.GOV.BR, 2016)

It can be seen from table 7 that in Silva Jardim the work order was not issued because of the debt with the TCE's Cadastre of Defaulters (Cadin), and the same is true of Itaocara and Pirai.

The service orders for the municipalities of Bom Jardim and Casimiro de Abreu have not yet been issued.

The cities of Paraiba do Sul, Santo Antônio de Pàdua, Sâo Francisco de Itabapoana, Sâo Joâo da Barra and Vassouras have not received executive projects.

Only the cities of Miguel Pereira and Paraty have their executive projects under analysis.

The best prospects for the project, as previously presented, are in the municipality of Engenheiro de Paulo de Frontin (Pilot Project). The last inspection was carried out in February 2016 and according to the Ministry of Communications' timetable, installation of the fiber optic cables and transmission equipment is expected to begin between March and April 2016. There will be nine access points, six for the government and three open to the public. The transmission towers will have a range of 300 meters. (PRESS OFFICE OF THE MAYOR OF PAULO DE FRONTIN, 2016)

Figure 4 shows and gives us an idea of the spatialization of the Program's status

Digital Cities bringing together all the selected cities in the state of Rio de Janeiro.

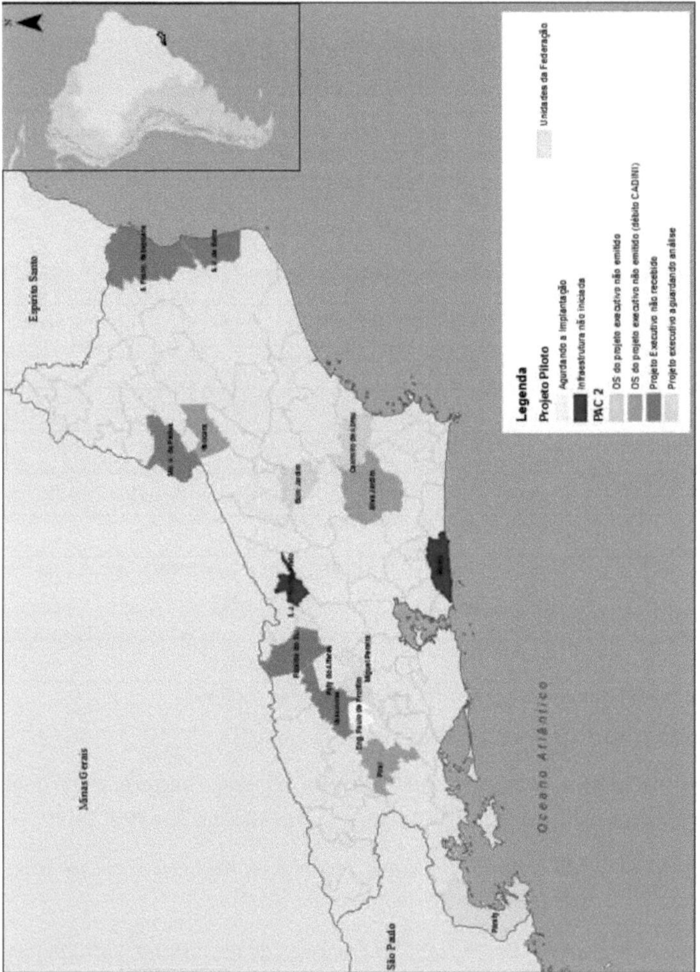

Figure 4- Project status of all the cities covered in the state of Rio de Janeiro Source: (CIDADES.DIGITAIS@COMUNICACOES.GOV.BR, 2016)

Rio de Janeiro is made up of 92 municipalities and according to IBGE data, the estimated population in 2015 corresponded to 16,550,024 inhabitants, leading us to conclude that the Digital Cities Program (Pilot Project + PAC2) would only serve 579,758 inhabitants, around 3% of the state's population. A very small number compared to the entire state.

2.3 - Third stage of the Digital Cities Program

From 2016 onwards, the program began to be restructured, with funding coming from parliamentary amendments to the budget assigned to the Ministry of Communications, under Action 00PA[6] . According to the MC, it will be carried out in partnership with Telecomunicações Brasileiras S.A (Telebrâs)[7] . This change was determined in ORDINANCE No. 4699, OF OCTOBER 14, 2015, which amends ORDINANCE No. 376, OF AUGUST 19, 2011, regarding the establishment of the Digital Cities Implementation and Maintenance Project.

The initial Digital Cities Project (2011) was amended, offering the following opportunities.

I - the implementation of connection infrastructure between local public bodies and equipment and to the Internet, including through the construction of backhaul[8] , in accordance with the specificities of each municipality and the administrative regions of the Federal District, promoting improvement and agility in the provision of services to citizens and the integration of public policies;

II - installation of public Internet access points for free use by the population;

Article 4 refers to management. That is: "it determines that the management and maintenance of the Digital Cities will be the responsibility of the municipalities served or the Federal District."

The technical feasibility study for the execution of the budget programs is the responsibility of Telebrâs and the Digital Inclusion Secretariat .[9]

In relation to the research's initial question, "Do these programs respond to the needs of citizens and

---

6 § Paragraph 3 - Resources from parliamentary amendments for investment that are earmarked for Budget Action 00PA (Participation of the Union in Capital - Telecomunicações Brasileiras S.A - Telebras - Implementation of Infrastructure for the Provision of Data Communication Services) may be used to carry out the actions provided for in this article.

7 See the full text of paragraphs 4 and 8 of ORDINANCE No. 4.699, OF OCTOBER 14, 2015

§ Paragraph 4 - Telebras and the Digital Inclusion Secretariat will carry out a study to verify the technical feasibility of executing the budget programs referred to in Paragraph 3.

§Paragraph 8 - The infrastructure deployed will be the property of Telebras, which may enter into partnerships with telecommunications companies and Internet access providers to ensure the expansion of access and the quality of the services provided to the population.

8 According to Teleco (2015) *Backhaul* can be defined as a connection support infrastructure for STFC (Fixed Switched Telephone Service) broadband, which interconnects access networks to the operator's *backbone*. It can be classified into

Wireless technology: through a cell site transmits data and voice to a *Switch* and from a central site to a remote one; Satellite technology: data is transmitted to a satellite via a point to which it can be transmitted (*Uplink*). It can also be used to transmit data to a network backbone.

9 The Digital Inclusion Secretariat (SID-MC) was created in May 2011 by Decree No. 7.It is responsible for: I - formulating and proposing policies, guidelines, objectives and goals relating to the Federal Government's digital inclusion; II - planning, coordinating, supervising and guiding the Federal Government's digital inclusion actions; and III - executing, following up, monitoring and evaluating the implementation of the Federal Government's Digital Inclusion Program, in conjunction with internal and external bodies and institutions. SID-MC's structure includes the Departments of Infrastructure for Digital Inclusion (DEID) and Articulation and Training (DEA).

local communities?" it can be seen that at the moment, at least the Digital Cities Program in Rio de Janeiro, does not give any indication of a positive response.

3. Expansion of the Digital Cities selected in Pac2 and pilots from other initiatives.

Building digital cities is a process of deepening and expanding reality itself (CINDIO; SCHULER, 2012). Infrastructure is (or should be, in this case) a basis for local, economic and social development. The Digital Cities Program - MC gives us clues that it cannot, at the moment, offer the expected transformations and benefits, because it is still in "latency".

Bearing in mind that the city is multidimensional, as Santos (2000) states, and is characterized as a place in which a mixture of more or less correct interpretations of the world, the country and the place itself is possible, other ways of analyzing digital initiatives in these municipalities become possible.

In spatial terms, the city is constituted as a large mesh of architectural and urbanistic systems that make sense of the physical and institutional infrastructure with horizontal and vertical density. This system creates life through the network of social relationships, which animate the city and drive human development towards material and even spiritual well-being.

Historically, the city is the place of encounter and disagreement, of the foreigner and the inhabitant. In this place, exchanges take place, culture takes shape, education and health are improved, food is mechanized, politics tries to make sense, laws are organized and justice is legitimized. It is also clear that it is here that violence, racism, fundamentalism and xenophobia are exacerbated, social exclusion takes its place and work ceases to be the instrument that frees human beings from ignorance and becomes a limited creative expression of the individual in the relationship and environment of employment and unemployment. In other words, it is in the city that we learn to develop and specialize our particularities and integrate them socially, but it is also where we discover the problems of our own lives as beings of relationships.

As a way of getting closer to the focus of our analysis, we should identify and map the potential of the cities themselves and their capacity to implement local development with technical and political flair. In doing so, we would be opening up the issue and promoting a constant debate, beyond analysis, with a view to finding alternatives and innovative solutions in city management, enabling systematic monitoring and a process of permanent evaluation of the social impacts of the actions implemented in the field of local, integrated and sustainable development. In other words, the city implies a specific system of social relations, culture and, above all, political institutions. The issue is broader if we explore local potential versus the material conditions for achieving local, integrated and sustainable development. However, the citizen cannot remain uninvolved or excluded from the benefits and potential of the information society and this is the challenge of technological management in cities.

The starting point in this field of technological convergence is the relationship between cities and the infinite possibilities opened up by technological advances and their application to local development, boosting the economic, social, cultural and political production of the place inhabited by the citizen and complementing the role of the real city, with its cyberspace dimension. The typology of the possibilities of digital cities reinforces the understanding that there are other forms of analysis since digital cities can arise spontaneously

3.1 - Types of Digital Cities

In the context of communication networks, the specializations and social function of Digital Cities will be defined by their managers, inductors and users. By way of example, some Digital Cities are characterized as cultural because they emphasize activities and dissemination in the field of culture; others as historical, because they preserve the history of their users; and some others have characteristics in the social field, and so on. There are endless examples and possibilities.

It is possible, however, to identify in the whole a great agility compared to a city.

Its creators are gifted and driven by creativity, even if this creativity comes in different stages. As the city develops, performance tends to improve.

The first digital cities emerge from the development of communication technologies and necessarily depend on a concentration of local infrastructures with technology nodes. This is the proposal of the Ministry of Communication's program.

Digital Cities emerge to meet the needs of a networked community, virtually created according to the needs of a real community, and are spread through a number of initiatives aimed at characterizing these virtual spaces. Depending on the promoters of these initiatives, we can characterize some Digital Cities. Among so many models, five types of Digital Cities will be defined (SILVA , 2004)

3.1.1 - Government Digital Cities are local or regional government initiatives.

In these cities, the government is usually the major service provider. They use information technology in public administration to provide online services, benefiting citizens by reducing bureaucracy and transparency in government functions, such as buying and selling services through digital auctions, certifications and others. In the current scenario, they are considered fundamental to decision-making processes in public and private organizations. Services can be differentiated depending on the municipality, state or country.

3.1.2 -- Non-governmental Digital Cities

Non-governmental Digital Cities are those in which digital communities can only enter after completing a registration form and obtaining a password to visit the city. Access can be charged or free. The organizing groups arouse the interest of the community by promoting digital leisure, meetings on social networks, providing access to various widely circulated magazines, encyclopedias, digital classifieds, tourist information, local information, etc.

3.1.3 -- The Digital Cities of Third Sector initiatives

These cities are made up of different organized social groups from civil society, such as Non-Governmental

Organizations, Foundations and Associations with the support of business and non-profit organizations.

They are usually thematic and deal with digital and social inclusion. They generate information on different topics, present social projects, fund projects or provide information on sources, among other things.

3.1.4 -- Spontaneous and individual initiatives

This city is different from the others and perhaps the most common. It is characterized by the spontaneous entry of the local digital community, albeit disorderly. Various groups and individuals are responsible for its architecture and organize themselves through certain links such as local websites, whether personal or not, social networks of the same name as the real city, frequenting electronic addresses that disseminate local news and events, freebies, other service-producing agents, consumers and those who practice the socialities of the local network. This dynamic is the result of social relations (flows) in the real city. As such, the fixed points are located in the Real City. In fact, this interpretation is close to the theoretical preposition of Milton Santos (1996, p.50), when he states that "fixed and flows together, interacting, express geographic reality and it is in this way that Conjuntamente appears as a possible object for Geography".

3.1.5 - Digital Cities of Mixed Initiatives

Digital Cities, in this case, are projects for digital and social inclusion, or services through partnerships between the public, private and third sectors.

Understanding the different concepts of a Digital City can be compared to a geography of cities, characterized and named by their function, conflicts, society, economy, etc. (SILVA , 2004).

3.2 Technological convergence of the various initiatives in these cities based on the Digital Cities models.

The local infrastructure to absorb digital cities, as we have already mentioned, is fundamental and for this we need to gather preliminary but sufficient information to understand the actions taken by communities, private and public initiatives.

The information society is defined, as discussed in the introductory document, Digital Cities in Brazil - White Paper (20101), by accessibility and connectivity using the World Wide Web as the means for establishing the production, transmission and social dissemination of information for public and private use.

Considering this premise, this society is organized based on the integration of the following key players:

- Public Power - expanding its field of universal services by electronic means and publicizing the mechanisms for citizen access;
- Private enterprise - boosting its capacity for business and commerce on the Internet;
- Third Sector - maximizing its power to exert political pressure and disseminate projects in the field of digital inclusion.

On the other hand, we must not lose sight of the fact that this society is consolidated and spread by government actions at all levels (municipal, state and federal), which therefore become essential for the universalization of access and digital inclusion, considering that, according to the principle of modernization of the state and the use of ICTs adopted by the government, the following premises are identified:

- The government is a service provider;
- Citizens are not concerned with details, but with the quality of public services and care;
- The municipality, with the help of the state and federal governments, is thus, in the context of the implementation, dissemination and consolidation of the information society, the main agent for inducing and promoting technological applications and universal access. However, these government properties imply, as we have already seen:
- Development of network infrastructure to absorb the demands of applying information and communications technologies;
- Implementation of local infrastructure for research & development, focusing on the local economic vocation and cultural identity;
- Articulation of partnerships in the field of educational development, with a view to generating a teaching and learning structure geared to market demands in the information society;
- Providing resources to enable local access to the benefits of the information society's potential;
- Stimulating and encouraging local business and e-commerce initiatives;
- Promoting and encouraging the formation of networks of information services that are strategically processed and consolidated with a view to the technological, economic, social and cultural development of citizens;
- Institutional reorganization and technological re-equipment aimed at offering faster public services focused on the citizen.

For effective transparency and a qualitative offer of these services, the government will need to emphasize the improvement of citizens' living conditions, which requires prioritizing:

- Staff training and citizen education;
- Standardizing the quality of services for citizens;
- Universalized and diversified accessibility based on situational demand;

- Fast connectivity with an ever-decreasing connection/person density;

When it comes to applying technology with a focus on human development, the City Hall also needs to create conditions for greater control and monitoring of the process of implanting and consolidating a society that benefits from the use of technology, implying the construction of an advanced system of digital communication and use of the internet aimed at the citizen. One of the solutions successfully experimented with is the creation, for example, of a public portal that directs information to the citizen.

Complementing this action, we can include the experience of public Internet access telecenters, which in this case offer..:

- Employee directory;
- Bureaucratic red tape and public services;
- Organizational structure of the City Hall;
- Reports on local/regional and national events;
- Institutional location maps for citizens;
- Other benefits of the right to information.

In short, a necessary and urgent modernization to the development and dissemination of new information and communication technologies.

It should also be borne in mind that modernization, with the potential that we see today, such as the use of cell phones, communication via email, Facebook, whatsup, has restructured response times and led to actions to improve the quality of services and the promptness with which citizens' requests are met.

In this sense, the research aims to identify some of the digital initiatives in the cities studied, the services offered to citizens through the municipalities' institutional websites and the cities' broadband technology bases.

What is the technological basis of the cities in the Program?

3.2.1 - Broadband

Broadband is the internet connection that allows users to surf at high speed. Simoes and Reis (2015) explain that the average broadband speed in Brazil is 3 Mbps and that the government wants to increase it to 25 Mbps by 2018.

Checking the predominant range of broadband access in the municipalities under study shows that the municipalities of Casimiro de Abreu, Sâo Francisco de Itabapoana and Silva Jardim are the highest ranked. In 2015, Brazil announced that it had 24.3 million fixed broadband access points. Of these, 46.3% are in the 2 Mbps to 12 Mbp range.

Table 7 shows the predominant speed range in each city and the access points. In Brazil, 3,971 cities have a predominant speed range of 512 Kbps to 2 Mbps.

Municipality(ies) PAC2	**Predominant range**	**Total access points**[10]
Bom Jardim	512kbps to 2Mbps	431
Casimiro de Abreu	2 to 12 Mbps	3437
Itaocara	512kbps to 2Mbps	2796
Miguel Pereira	512kbps to 2Mbps	3525
Paraiba do Sul	512kbps to 2Mbps	3067
Paraty	512kbps to 2Mbps	3025
Pirai	512kbps to 2Mbps	1800
St. Anthony of Pàdua	512kbps to 2Mbps	2239
Sâo Francisco de Itabapoana	2 to 12Mbps	1194
Sâo Joâo da Barra	512kbps to 2Mbps	2067
Silva Jardim	2 to 12Mbps	778
Brooms	512kbps to 2Mbps	4229

Source: Anatel (2015)

Municipality(ies) Pilot Project	**Predominant range**	**Total access points**
Engenheiro Paulo de Frontin	512kbps to 2Mbps	1035
Marica	512kbps to 2Mbps	9130
Sâo José do Vale do Rio Preto	512kbps to 2Mbps	2067

Source: Anatel (2015)

With regard to 4G coverage[11] in Brazilian municipalities, 477 municipalities have it and it serves 55.1% of the population. Of the cities studied, 11 have some kind of 4G coverage: Bom Jesus do Itabapoana, Casimiro de Abreu, Engenheiro Paulo de Frontin, Itaocara, Maricà, Miguel Pereira,

10 A network connection that allows a computer or user to connect to a corporate network. Virtual private networks (VPNs), wireless communications and Remote Access Service (RAS) dial-up connections are examples of access points.

11 4G is the acronym for the Fourth Generation of mobile telephony. 4G is based entirely on IP, being a system and a network, achieving convergence between cable and wireless networks and computers, electronic devices and information technologies to provide access speeds of between 100 Megabit/s on the move and 1 Gigabit/s at rest, maintaining an end-to-end (point-to-point) quality of service (QoS) with high security to make it possible to offer services of any kind, at any time and in any place (citation out of date). (SEIDEL , 2015)

Parati, Pirai and Santo Antônio de Pàdua according to Table 8 - List of municipalities covered by 4G, the other cities Bom Jardim, Paraiba do Sul and Silva Jardim do not have 4G coverage to date.

Cities	Live	TIM	Of course	Hello	Total
Bom Jesus do Itabapoana	-	X	-	-	1
Casimiro de Abreu	-	X	-	-	1
Engenheiro Paulo de Frontin	-	X	-	-	1
Itaocara	-	X	-	-	1
Maricà	-	X	X	-	2
Miguel Pereira	-	X	-	-	1
Parati	X	X	X	-	3
Pirai	-	X	-	-	1
St. Anthony of Pàdua	-	X	-	-	1
Sâo Francisco de Itabapoana	-	X	-	-	1
Sâo Joâo da Barra	-	X	-	-	1
Brooms	-	X	-	-	1

Table 8: List of municipalities covered by 4G Source: Teleco (2016)

It can be concluded that Bom Jardim, Itaocara, Miguel Pereira, Paraiba do Sul, Paraty, Pirai, Santo Antônio de Pàdua, Sâo Joâo da Barra, Vassouras, Engenheiro Paulo de Frontin, Maricà and Sâo José do Vale do Rio Preto are in the national average.

The prevailing speed range in each city defines the speed of the demonstrations in the digital city.

3.2.2 Community Internet Program

The state of Rio de Janeiro has a Community Internet Program (CIC) developed by PRODERJ (Center for Information and Communication Technology of the State of Rio de Janeiro), which implements computer labs offering the population free digital literacy training, broadband web access and various e-government services. (COMMUNITY INTERNET, 2016)

Users of the CICs are guided and can carry out various services. Such as: Requesting 2ª copies of bills (Light and Cedae, Creating e-mails, Printing and sending CVs, Registering for competitions, Pre-enrolment at school, Carrying out school surveys, Consultation, by CPF, of income tax refunds already released, Detran/RJ services, Access to various INSS services, such as scheduling medical examinations;

requesting an extension of leave; requesting maternity allowance, among others. Citizens can take part in specific digital literacy training, learning basic computer skills and how to navigate the web (INTERNET COMMUNITARIA, 2016). However, the user is given a time limit of 30 minutes per session and depending on the activities to be carried out, this time may not be enough. You can remain at the machine if there is no queue according to the access rules.

The program is offered in several cities in Rio de Janeiro, but in relation to the cities studied, it is present in: Casimiro de Abreu (two), Engenheiro Paulo de Frontin (three), Miguel Pereira (one), Paraty (one), Sâo José do Vale do Rio Preto (one), and Vassouras (one), according to Figure 5. Program cities with Proderj Community Internet Centers (CICs)

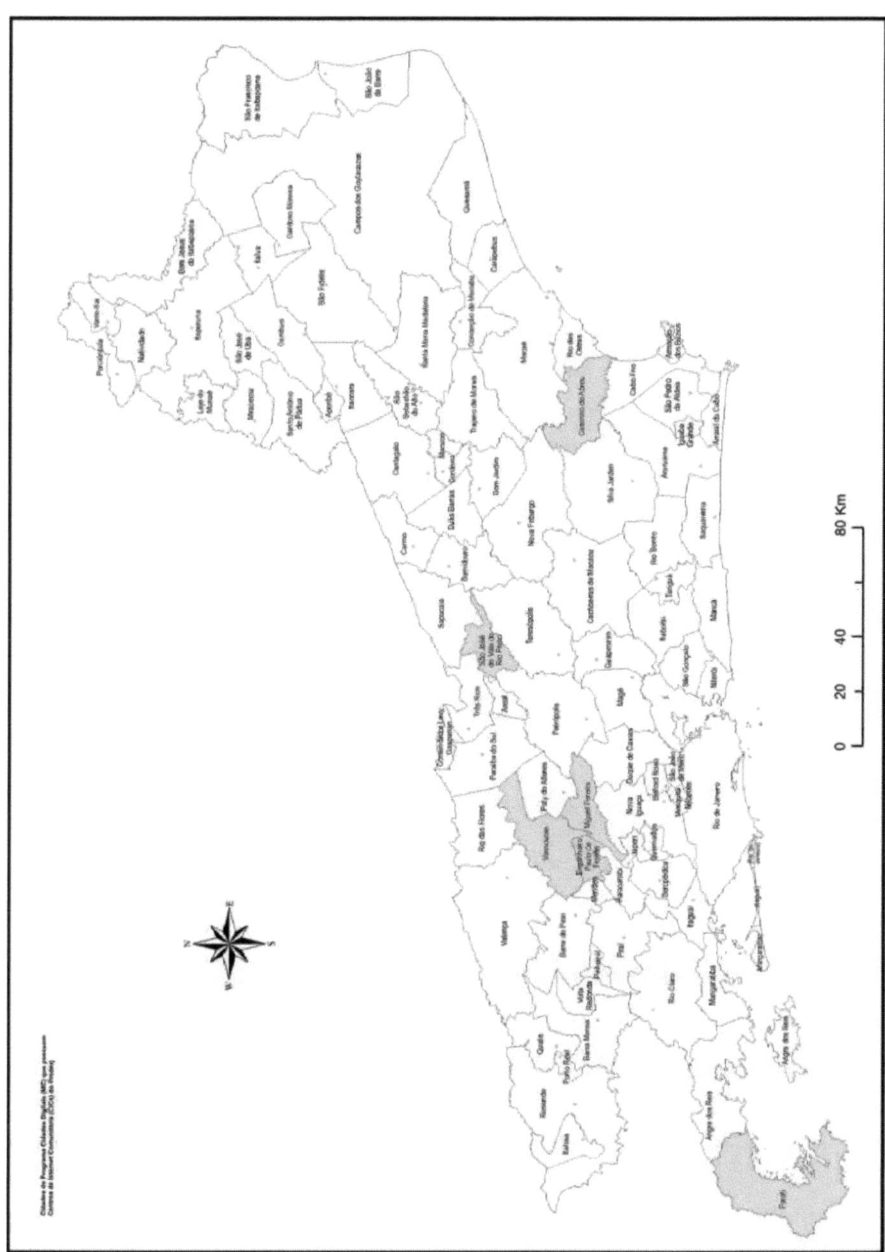

Figure 5 - Program cities with Proderj Community Internet Centers (CICs)

Source: CIC (2016)

3.2.3 . Modernization services for town halls, identified on official websites.

As we have already mentioned, it is possible to study the basis of sociability relations in the virtuality of cyberspace and its reflexes on the material basis of society, as well as to create a collection of information in the most varied forms and easily accessible about the reference city. With this in mind, the research looked at the main services offered on the town hall websites and some local initiatives in the field of information and communication technology. The aim is to understand how technology can help by facilitating opportunities for citizens and at the same time gain control over various issues, such as municipal transparency, which directly involve the city's financial health, remembering that the digital city is much more than a website.

a) Casimiro de Abreu:

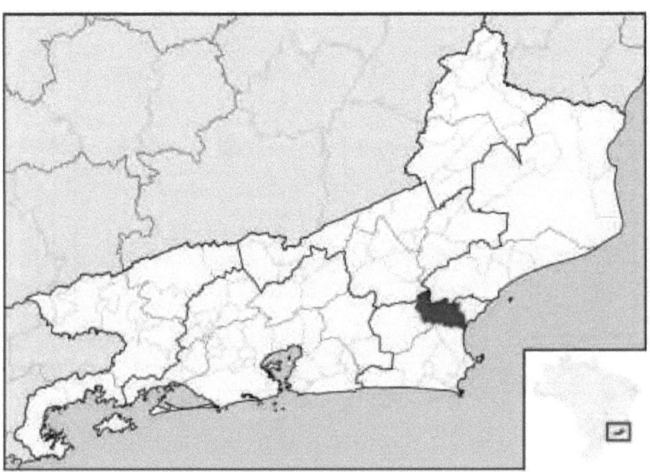

Source: Raphael Lorenzeto de Abreu (2016)

The main services offered on the Casimiro de Abreu City Hall website	
Transparency portal	Income, Expenditure, Assets, Warehouse, Personnel, Purchasing/Tendering, Statements, Budget, Access to Information.
The BUY MORE CASIMIRO DE ABREU Project,	Its aim is to modernize the entire system of government procurement carried out by the municipality. Through tools and facilitating mechanisms, our goal is to bring companies ever closer to the purchasing body, training all the people involved in the purchasing process, be they civil servants or businesspeople, so that the two are better integrated. In this way, we seek a better advantage for the Public Administration, as efficiently and effectively as possible.(http://www.com prascasim iro.com

	.br/About)
Ombudsman	The Ombudsman's Office is a channel for dialogue between the municipal government and the public. It is an open door for public participation, aimed at exercising citizenship and improving the quality of the public services provided. (http://www.casimirodeabreu.rj.gov.br/ouvidoria.html)
Email Services	Email exchange services
Legislation	Main laws, decrees and codes of the municipality. Provides for the reformulation of the administrative structure of Casimiro de Abreu City Hall
Garbage collection	Presentation of the itinerary
E-sic	Public Administration Information Channel
Process Consultation	You can consult various processes
IPTU - Your IPTU makes a difference	2ª copy of IPTU
Notices and Calls	Public notices for City opportunities available
Electronic Invoice	Issue invoices electronically
Education and Health	Online teacher, nursery registration, school units, timetables, school bulletins, health units and more...

Table 9 - The main services offered on the Casimiro de Abreu City Hall website Source: Casimiro de Abreu website (2016)

Local initiatives in the field of information and communication technology in Casimiro de Abreu

- The Conecta Casimiro de Abreu Project offers free wireless signal in five parts of the municipality and its districts.

The Community Internet Center in Casimiro de Abreu (there are two centers)

b) Itaocara:

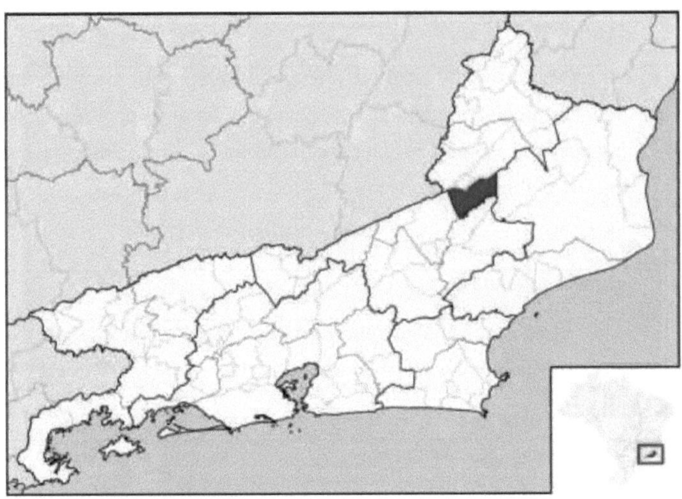

Source: Raphael Lorenzeto de Abreu (2016)

The main services offered on the Itaocara City Hall website	
Citizen Space	Access Request, Certificates, Authenticity Consultation, Forgot Password, 2nd Copy of Receipt, Permit, Collection Guide and Contact Us
Pronim Transfer Brazil	Income, Expenses, People Management, Creditors, Assets and Administration.
Electronic Invoice	Issue invoices electronically
Online payslip	For servers
DEISS	Electronic ISS declaration
Ombudsman of the City Hall and the Public Hospital	The Ombudsman's Office is a channel for dialogue between the municipal government and the public. It is an open door for public participation, aimed at exercising citizenship and improving the quality of the public services provided. (http://www. casimirodeabreu. rj.gov.br/ouvidoria. htm l)

Table 10 - The main services offered on the City Hall website Source: Itaocara website (2016)

Local initiatives in the field of information and communication technology in Itaocara

No projects were found that reveal digital inclusion, and the site offers few online services.

c) Miguel Pereira

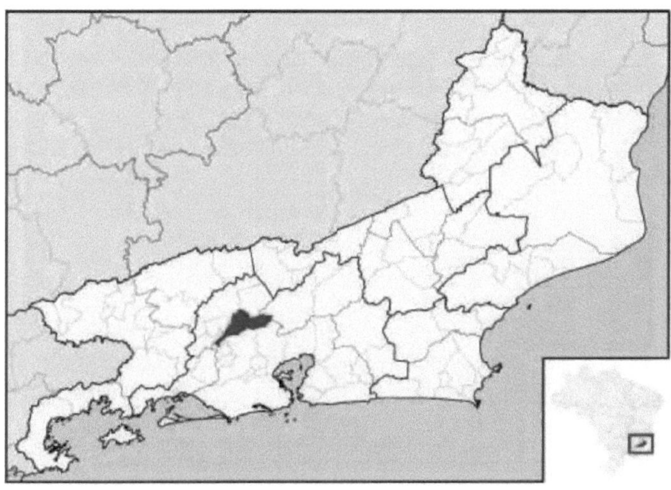

Source: Raphael Lorenzeto de Abreu (2016)

The main services offered on the Miguel Pereira City Hall website	
Gate of transparency	Tenders,
BIN	Municipal newsletter
Access to information	A space for participation (Participate), Services (Fies, Prouni, etc.), Legislation and Channels
CEPIM	CEPIM's records are based on the information entered into the INTEGRATED FINANCIAL ADMINISTRATION SYSTEM - SIAFI by the federal public administration bodies and entities that grant funds. Private non-profit organizations with "EFFECTIVE DEFAULT" or "IMPUGNED" registrations are considered to be barred. Therefore, any clarifications or requests to correct CEPIM's information should be made to the aforementioned bodies and entities. The data published on CEPIM is updated daily.
SIGFIS	Sending to the Federal Court of Auditors

Table 11 - The main services offered on the City Hall website Source: Miguel Pereira website (2016)

Local initiatives in the field of information and communication technology in Miguel Pereira

- Community Internet Center in Miguel Pereira (one center)

d) Paraiba do Sul

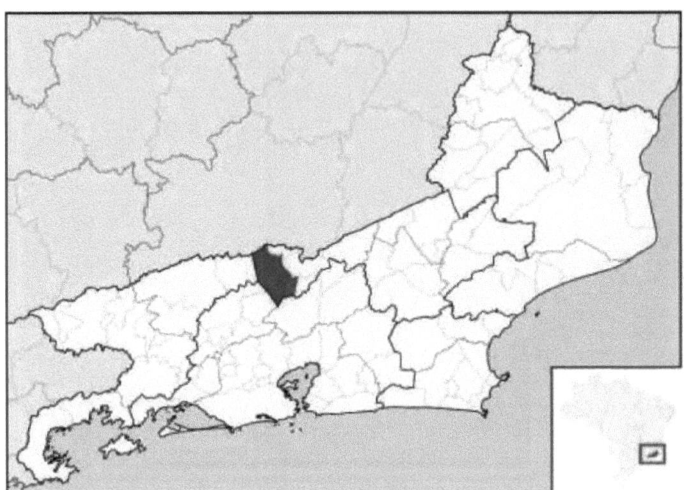

Source: Raphael Lorenzeto de Abreu (2016)

The main services offered on the Paraiba do Sul City Hall website	
Transparency Portal	It was created with the aim of promoting broad and objective access to data on the use of municipal public resources. Through it, citizens can monitor the management of the finances of all the bodies and entities that make up the Direct and Indirect Administration.

Table 11 - The main services offered on the City Hall website

Source: Paraiba do Sul website (2016)

Local initiatives in the field of information and communication technology

No projects were found that reveal digital inclusion, and the site offers few online services.

e) Paraty

Paraty has some modernization services from the Citizen Web site

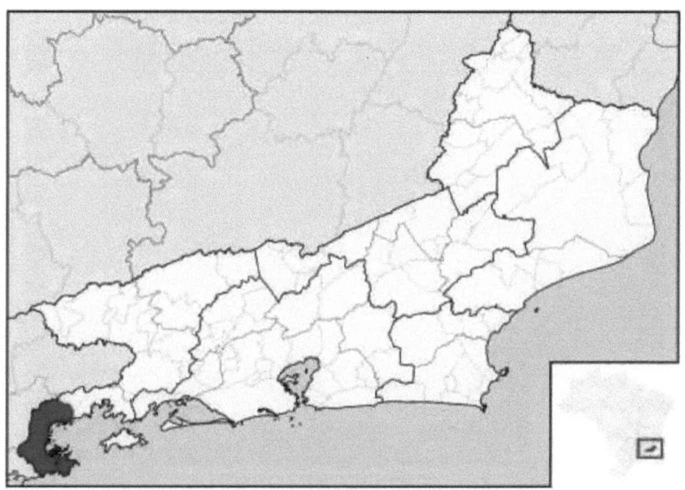

Source: Raphael Lorenzeto de Abreu (2016)

The main services offered on the Paraty City Hall website.	
Consult	Consults taxpayers and validates documents
Taxpayer details	Change of registration data.
Issue	Negative Taxpayer Certificate, Negative Property Certificate, Negative Economic Certificate, Negative Property Certificate, ITBI Certificate, 2ª copy of the Exemption Certificate and ISS Guide
Service Request	Identification as a taxpayer or company is required
Legislation	Main laws, decrees and codes of the municipality. Provides for the reformulation of the administrative structure of the City Hall
Electronic Invoice	Issue invoices electronically

Table 12 - The main services offered on the City Hall website Source: Paraty website (2016)

Local initiatives in the field of information and communication technology in Paraty

- **Community Internet Center in Paraty (one center)**

Project "Conexâo 3ª Idade: inclusâo digital" carried out at the Centro de Internet Comunitària/Proderj:

- **Digital whiteboard**

With the aim of improving and innovating teaching in Paraty's municipal education network, a total of 14 schools benefited from the initiative.

- **There are several Telecenters, one example of which is the Pesca-Maré one**, aimed at fishermen in the Costa Verde region. The space has 10 computers connected to the internet with a multimedia kit involving a printer, projection screen, data show and home theater

f) Pirai

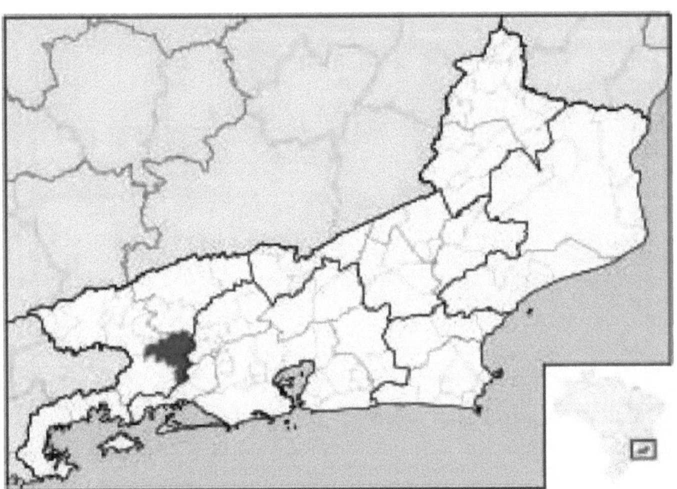

Source: Raphael Lorenzeto de Abreu (2016)

The main services offered on the Pirai City Hall website (Pirai 2.0)[12]

Laws	Main laws, decrees and codes of the municipality. Provides for the reformulation of the administrative structure of the City Hall
Transparency	Income, Expenses, Assets, Warehouse, Purchasing/Tendering, Personnel, Statements, Budget, Access to Information.
e-SIC)	The Electronic System for the Citizen Information Service (e-SIC) allows any person, physical or legal, to submit requests for access to information, track the deadline and receive a response to the request made to agencies and entities of the Federal Executive. Citizens can also file appeals and complaints without bureaucracy.
Contracts and Tenders	Electronic Auction
Ombudsman	The Ombudsman's Office is a channel for dialogue between

12 The Piraí 2.0 Portal can also be accessed via cell phone, whose app is available on Google Play,

	the municipal government and the public. It is an open door for public participation, aimed at exercising citizenship and improving the quality of the public services provided.
IPTU Guide Issuance	From the information of the CNPJ/CPF or the Cadastral Inscription of the Property
Services	Requests for public lighting, online paychecks for civil servants, and a business guide full of information on companies in the municipality, whether in the commercial, service or tourism sectors.

Table 13 - The main services offered on the City Hall website

Local initiatives in the field of information and communication technology in Pirai

Pirai has already started the digital city process since 2000. This vision of the city of the future is a response to the dismissal of 1,200 Light employees as a result of the neoliberal policy of privatizing state-owned companies. The impact on the social, economic and political spheres led the municipal administration to look for new opportunities using technology in favor of local development.

The year 2004 was a milestone as the municipality's digital culture began to be disseminated, "involving digital inclusion actions, education for new media and computerization of management" (PIRAI DIGITAL, 2015).

According to the promoters of Pirai digital, the results can already be seen:

> Tax collection and other credits were increased; Tax, Financial and Administrative structures were integrated; Inspection became more efficient and effective; Information and services were made available immediately and accurately; Servants were trained and became able to absorb new technologies; Service bureaucracy was reduced; The City Council's relationship with the population became interactive; The efficiency of service implementation was guaranteed; Transparency in administration was increased; (PIRAI DIGITAL, 2015).

Pirai also acquired the UNESCO seal of approval in 20014, from the Pirai Digital and Local Development Project for its initiative to democratize access to information and communication media through new technologies. Another significant award in Latin America is the "Latin American Digital Cities Award" from AHCIET (Asociacion Iberoamericana de Centros de Investigacion y Empresas de Telecomunicaciones).

The project reveals that at first the connection infrastructure was wireless, but at a very high cost. In the second phase, it was optimized with a cable connection between buildings. The network is currently configured as follows:

> The current coverage of the SHSW Network: the entire municipality of Pirai, covering 39 public buildings with 144 computers, plus 20 educational establishments with fast internet, for a total of 6,300 students and 188 computers (33 students per computer).

Access to the SHSW network also includes 20 buildings with 66 computers, covering four libraries, a Children's Home, an APAE, a Nursery School, four Telecenters and nine Kiosks. This represents a total of 398 computers available not only to students, but to the entire community.

The Telecenters already serve 220 people a day (students, housewives, teachers and university students), for purposes such as recipes, bank accounts, research and entertainment. (PIRAI DIGITAL, 2015)

According to data from the Pirai Digital website (2016), the Pirai Digital Network has 107 Connected Points, 23 Neighborhoods served with Internet, eight Squares with Internet, 3160 Registrations with approximately 12,640 people benefiting from free Internet at home, and in 2013 there were 970 registrations.

From the perspective of Pirai Digital, the theme was divided into .ORG, .GOV, .EDU and

.COM.

- .ORG - According to the project, the aim is to share with the community ways of incorporating new information and knowledge technologies into people's daily lives, transforming citizens into the main actors in producing, managing and enjoying the benefits of new information and communication technologies.

- The results of .org - Telecenters have been installed in every district, Public access terminals have been placed in every neighborhood, The number of people using basic software and accessing the Internet has increased, The specific skills that information and communication technology enables have increased, The population uses terminals for consultations, complaints and participation.

- .GOV, its aim is to innovate and transform public administration based on the principles of trust and transparency, excellence in service, participation and cooperation and technological qualification through commitment to modern information technologies.

- .EDU - its objective is to contribute to the development of competencies and skills for the formation of a citizen capable of living together, communicating and dialoguing in an interactive and interdependent world using information and communication technologies in innovative pedagogical and educational activities.

- .COM - the aim is to integrate companies in the creation of a telecommunications infrastructure network as a community asset from the perspective of social responsibility in the process of digital inclusion. The results have made it possible to raise financial and infrastructure resources, the network has been maintained and extended, and both the community and the company have access to the network.

Pirai is also on the social networks youtube, Twitter, Facebook and has a program on TED[13] called TEDx Pirai , a program of local, self-organized events that bring people together to share a TED-style experience.

You can also see the ELL and X-Cross projects

- The ELL project is an international and innovative program that involves students, educators and researchers and aims to teach the English language as a tool to make cultural creation concrete and meaningful, helping children's social and cultural relationships in elementary school, enabling a more solid intellectual development for the student.
- The X-Cross Project is a multimedia learning project through which 15 high schools in Latin America and Europe can form part of a transcontinental network. According to the project, the participating cities are: Miraflores (Peru), Pirai and Vitória (Brazil), Segrate (Italy) and the city of Bremen (Germany). The main objective of this project is to make information and communication technologies essential components for instruction and learning on both continents.

The City of Pirai is increasingly structuring itself to offer more digital services to the community and is already a world reference. We understand that the Digital City is a process that improves as new technologies are inserted into the geographical space, in other words, it doesn't end.

g) Santo Antônio de Pàdua

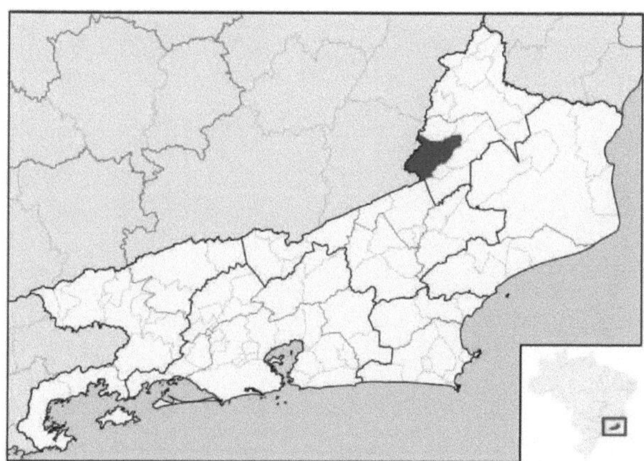

Source: Raphael Lorenzeto de Abreu (2016)

The main services offered on the Santo Antônio de Padua City Hall website

13 (A digital conference program that accumulates lectures on topics with a lot of repercussions)

Useful links	ISS - DEISS, Online Citizen, Sanitary License Model, Online Counter Check, Public Tender, Application Models, Organic Law, Code of Posture, Code of Works and Buildings, Electronic Invoice
Talking to the Mayor	Send and follow the progress of the process via the web or through our android app.
Online payslip	Information for civil servants
Transparency Portal	Social Assistance, Acts, CMAS, CMDCA, TUTELAR COUNCIL, Agreements, Decrees, Notices, FAP, Payroll, Laws, LICENSE, BIDDING, Others, Ordinances, Fiscal Responsibility, Review, PPA- 2015/2017, Secretaria . Municipal Health Department, Education Department, Contracts, Suppliers, Creditor Bids, Income and Expenditure.

Table 14 - The main services offered on the City Hall website

Source: Pairai website (2016)

Local initiatives in the field of information and communication technology in Santo Antônio de Pàdua

No projects were found that reveal digital inclusion, and the site offers few online services.

h) São Francisco de Itabapoana

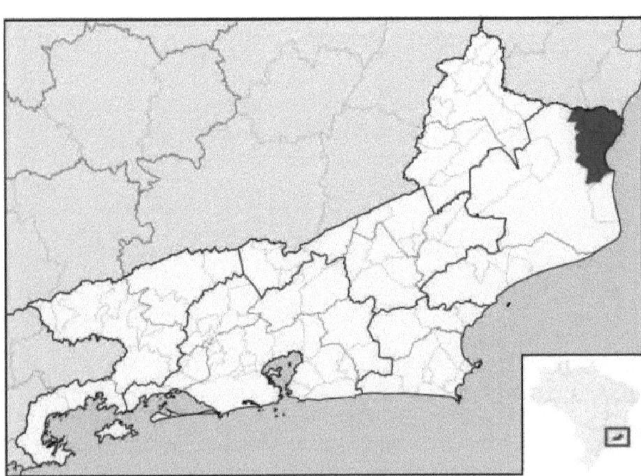

Source: Raphael Lorenzeto de Abreu (2016)

The main services offered on the website[14] of the Municipality of Sao Francisco de Itabapoana	
Transparency portal	**Budgetary Planning:** PPA - Multiannual Plan, LDO - Budget Guidelines Law, LOA - Annual Budget Law. **Budget Execution:** Income, Expenditure, Transfers, Reports. **Presentation of Accounts:** Balance Sheets and Reports **Fiscal Responsibility:** RREO, RGF. **IN-28 (TCU):** Reports
SFI Digital (under maintenance)	Database of legal norms
Downloads (in maintenance)	Summons, Miscellaneous, Notices, Legislation and others
SIA - Collection System	Real estate, furniture, taxpayers and ITBI
Electronic Invoice	Issue invoices electronically
Job and income opportunities	Companies provide jobs

Table 15 - The main services offered on the City Hall website

Source: Sâo Francisco de labapoana website (2016)

Local initiatives in the field of information and communication technology in Sao Francisco de Itabapoana

- Barrinha's digital inclusion project, inaugurated in 2012 in partnership with the Sâo Francisco de Itabapoana City Council, SERPRO-RJ (Rio de Janeiro Federal Data Processing Service), UENF (Darcy Ribeiro North Fluminense State University) and the quilombola community's board of directors. There are around 860 quilombolas with access to 77 computers.

- Community Internet Center in Sâo Francisco de Itabapoana (one center)

i) Sâo Joâo da arra

14 The site displays a warning that it is being redesigned (2016)

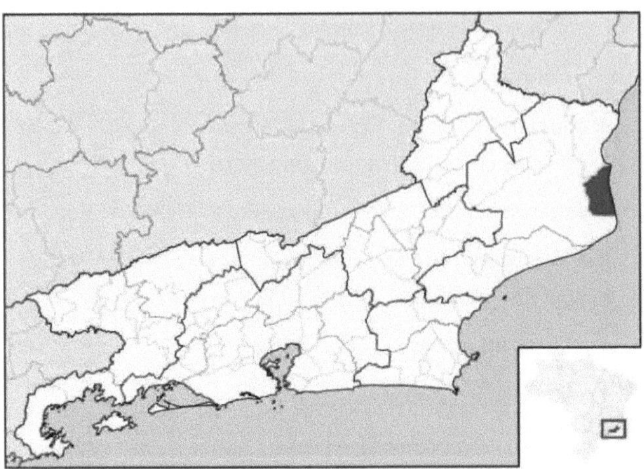

Source: Raphael Lorenzeto de Abreu (2016)

The main services offered on the Sao Joao da Barra City Hall website	
E-SIC	Electronic system of the Federal Government's citizen information service linked on the page
Government	- Secretariats, - Transparency, Tenders, - Master Plan and Official Gazette
Citizen	IPTU, - Useful Phones - Legal Assistance - Jobs Directory - Bus Timetable - Taxi Stations - Health Centers
Server	- payslip - payment schedule - staff regulations - position/salary plan - webmail
Tourist	- Beaches and Lagoons - Natural Attractions - Entertainment and Leisure - Tourist Calendar - Inns - Restaurants - Pharmacy
Electronic Invoice	Issue invoices electronically

Table 16 - The main services offered on the City Hall website Source: Sâo Joâo da Barra website (2016)

Local initiatives in the field of information and communication technology in Sao Joao da Barra

- Computer Lab - Set up at the headquarters of the Z-02 Fishermen's Colony in Atafona, in São Joâo da Barra. The aim is to contribute to the digital inclusion of fishermen in this region.
- Computer Laboratory at the Advanced Center of the Fluminense Federal Institute (IFF) and the digital inclusion project "Radical Digital City" aim to develop information technology infrastructure and services in the municipality. The city

also an Advanced Center of the Fluminense Federal Institute of Education, Science and Technology.

- City squares - Free internet access, according to Sao Joao da Barra City Hall, hotspot system and antennas with greater capacity allow access in public places.

j) Silva Jardim

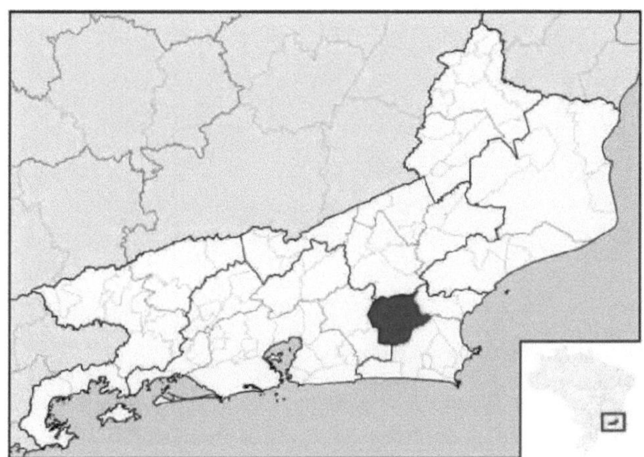

Source: Raphael Lorenzeto de Abreu (2016)

The main services offered on the Silva Jardim City Hall website

City Hall	The Municipality, Mayor, Deputy Mayor, Secretariats, Competitions, E-SIC (Ombudsman), E-SIC (Ombudsman) - ANSWERS.
Legislation	Decrees, Public Accounts, Tenders, Tender Notices, Price Registration Minutes, Contracts, Official Bulletin, Transparency Portal, Comptroller's Office - CGM
Services	IPTU, Nota Fiscal, Contra Cheque, Web Mail, Web Mail (New)
WEB TV Silva Jardim	A TC with news from the city can be accessed on the website.

Table 17 - The main services offered on the City Hall website Source: Silva Jardim website (2016)

Local initiatives in the field of information and communication technology in Silva Jardim

No significant digital inclusion policies found

l) Brooms

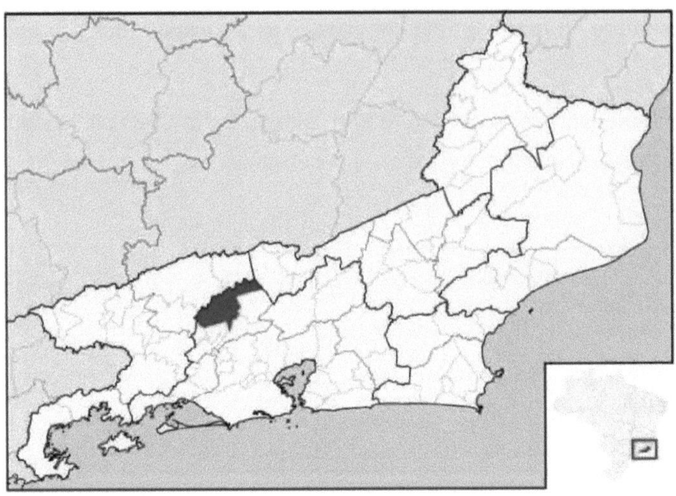

Source: Raphael Lorenzeto de Abreu (2016)

The main services offered on the Vassouras City Hall website

City Hall	Centro da Cidadania - a space for providing services and activities aimed at the well-being of civil servants and taxpayers in a single location. In addition to the main municipal departments, the Citizenship Center brings together services from different state and federal agencies.
Transparency Gate	Online Transparency System for online publication of records of competencies and organizational structure, addresses and telephone numbers of the respective units and public service hours; records of any transfers or transfers of financial resources; records of expenditure; information relating to tenders, including notices and results, and all contracts entered into; general data for monitoring programs, actions, projects and works of bodies and entities.
Online Accountability	Full-time publication of the City Council's accounts

Table 18 - The main services offered on the City Hall website Source: Vassouras website (2016)

Local initiatives in the field of information and communication technology in Vassouras

- FAETEC's Center for Digital Democratization (CDD)

Faetec Digital, a unit enabling the local population to have unlimited access to free broadband internet. (PORTAL DA CIDADE, 2015)

- Digital Literacy Course in the Cambota neighborhood, the Valença Rural Producers' Union plans to carry out the project in communities across the municipality. The fast-track course, which lasts four days, teaches basic computer skills to people in need, with a focus on senior citizens and children. (PORTAL DA CIDADE, 2015)

- Community Internet Center (CIC)

Digital literacy courses, offered free of charge by PRODERJ, help include the elderly in the virtual world.

m) São José do Vale do Rio Preto

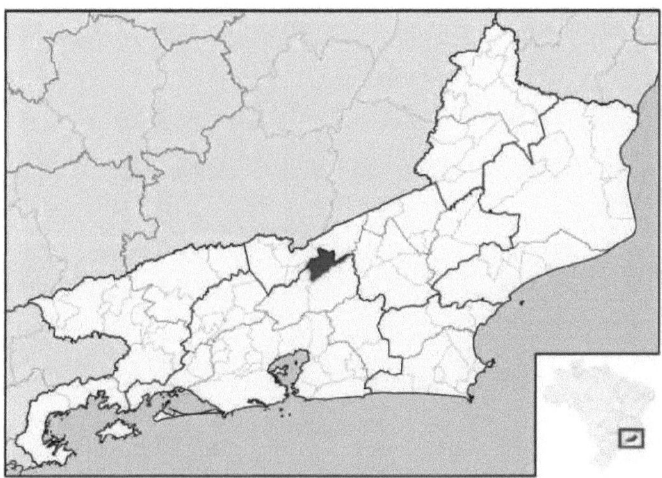

Source: Raphael Lorenzeto de Abreu (2016)

The main services offered on the São José do Vale do Rio Preto City Hall website	
Official Gazette	Provides the official gazette for monitoring the acts of the municipality.
Transparency	Basically, it presents activity reports (purchase reports, seedling donations, meal reports and others).
Tenders	It's a hyperlink that informs you that no items have been found.
Tender results	The name is self explanatory: it shows the results of online trading.
Tax incentives	For example, product exemptions, tax and administrative treatment.
Municipal legislation	Dissemination of laws, decrees, supplementary laws and ordinances.
Suppliers	There is no information in this section of the portal, only the link

Electronic invoice	No service despite the link

Table 19 - The main services offered on the City Hall website

Source: São José do Vale do Rio Preto website (2016)

Local initiatives in the field of information and communication technology in Sao José do Vale do Rio Preto

- Community Internet Center (CIC)

Digital literacy courses, offered free of charge by PRODERJ, help to include elderly people in the virtual world.

n) Maricà

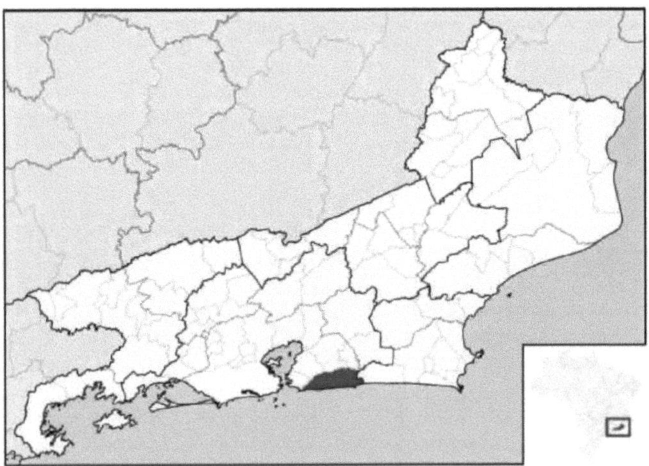

Source: Raphael Lorenzeto de Abreu (2016)

The main services offered at the Maricà site	
Transparency portal	Expenses, income, per diems, payroll, tenders, contacts and other information.
e-SIC	Electronic system of the Citizen Information Service that allows any person, physical or juridical, to send requests for access to information to bodies and entities of the Municipality of Maricà.
Ombudsman	You have the option of making **an Anonymous Complaint.**
JOM	Maricà's official newspaper
NFS-E	The NFS-e Electronic Service Invoice Portal allows you to automate the ISS payment form.

Company + Easy	Any entrepreneur can formalize their business. You can request a feasibility study
Purchasing Sector	The Purchasing Department is responsible for establishing the flow of materials, following up with the supplier and speeding up delivery. In this section of the website you will find: SRP - Price Registration System and the Suppliers' Area.
IPTU ONLINE	Withdrawal of the 2016 IPTU and also for Divida Ativa to request a debit sheet, payment slip or other information.
GISS Online	Giss Online - according to the link WHAT IT IS, explains that its objective is to enable knowledge of the Economy of the Municipality in the Services Sector in all its peculiarities, providing macro and micro economic data and information in real time, thus enabling the Public Administrator to make a real analysis of Fiscal Economic behavior. It features an exclusive registration area for accountants, condominiums, condominium management, companies not established in the municipality and others.
Server Portal	The aim is to find all the services and information that will help you in your daily life.
Health	List of health centers, opening hours.
Smart Enrolment	Streamlines the enrollment process in the municipal school network

Table 20 - The main services offered on the City Hall website

Source: Maricà website (2016)

Local initiatives in the field of information and communication technology in Maricà

- Maricà has the Marica Digital - Smart City Program. The project is a process of democratizing the Internet. The aim is to progressively expand it in order to reach as many residents as possible. According to the website, through this initiative, the internet signal is reaching the districts of Itaipuaçu, Inoa and Centro free of charge. The distribution of the free internet signal is done by radio, within a wireless network, Wi-Fi (with a frequency of 2.4 Ghz). Access is provided from within the area covered by the signal distribution or from any public place where the existence of the wireless network can be identified (MARICA, 2016).

- Public Access **Points**:

- Orlando de Barros Pimentel Square,
- Conselheiro Macedo Soares Square
- Maricà City Hall
- USF Jardim Atlântico 1 and 2
- Marqués de Maricà Municipal School
- Marcus Vinicius Municipal School
- CAIC Municipal School
- Barra de Zacarias Municipal School
- Tatiana Chagas Memory Municipal School
- Taques Horta Municipal School

- **Maricà has different networks available to the population free of charge. Table 8 shows the networks in the neighborhoods covered.**

Center	Itaipuaçu	Inoa
City Hall_Net_00 City Hall_Net_01 City Hall_Net_02 City Hall_Net_03 City Hall_Net_04 City Hall_Net_20	City Hall_Net_05 City Hall_Net_06 City Hall_Net_07 City Hall_Net_12	City Hall_Net_08
Nook	**Santa Paula**	**Sao José**
City Hall_Net_09	City Hall_Net_R	City Hall_Net_10 City Hall_Net_11 City Hall_Net_13

Table 21 - Networks of the neighborhoods covered Source: Smart City of Marica (2016)

- **Community Center for Social Assistance of the Legion of Good Will.**

Digital inclusion projects based on the computer courses offered. It teaches various computer tools to girls and boys enrolled in the **Criança: Futuro no Presente! (Child: Future in the Present!)** program.

- **Digital Home.**

It offers Digital Inclusion courses for senior citizens.

- **Distribution of laptops to municipal students**

o) Engenheiro Paulo de Frontin -

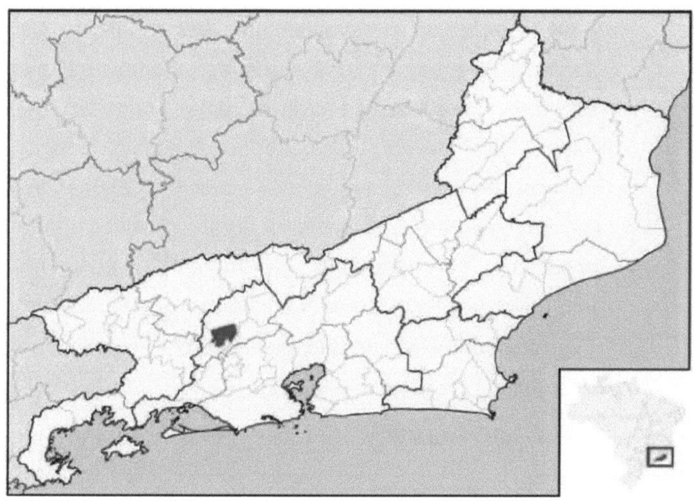

Source: Raphael Lorenzeto de Abreu (2016)

The main services offered on the Engenheiro Paulo de Frontin website

Transparency portal	• Revenue - Funds received by the government municipal • Expenses - The site has the following links: Expenditure by creditor, Execution of expenditure, Execution of programs, Direct government expenditure, Direct expenditure by expenditure, Direct expenditure by beneficiary, Direct expenditure by body, Direct expenditure by project/activity, Financial transfers to third parties, Committed expenditure, Commitments payable in chronological order, Expenditure on travel expenses, Expenditure on travel expenses by creditor, Expenditure on corporate cards and fund supplies. • Tenders - Follow-up queries of society - Law 131/09.
	• Purchases - Here you can see the list of purchases made by the municipality. • Staff - The site has the following links:

	Staff, Active civil servants/employees, Inactive civil servants, Permanent civil servants, Contracted civil servants, Commissioned civil servants, Trainees, Total payroll, Salary levels, Transferred/received civil servants, Political agents, Self-employed people Statements - (accompany the statements of funds received and/or spent by the municipal government) Agreements, Intragovernmental financial transfers, Voluntary transfers, Extra-budgetary income and expenditure. - Access to Information: Register a request, Order query and FAQs.
E-Note	Issue electronic invoices
Ombudsman	Citizens can send in their requests, questions, suggestions, complaints or any other type of manifestation.

Table 22 - The main services offered on the City Hall website Source: Engenheiro Paulo de Frontin website (2016)

Local initiatives in the field of information and communication technology in Engenheiro Paulo de Frontin.

- **liberated WIFI Project -**

The Board of Directors of the City Council provides free access to the internet through the network WI-FI .(CAMARA MUNICIPALI, 2015)

- **Community Internet Program (CIC)**

Engenheiro Paulo de Frontin I

Engenheiro Paulo de Frontin II

Engenheiro Paulo de Frontin III

4. Final considerations

Perhaps the best expression of the challenge facing the construction of the Digital City is to create an environment in the city that is frankly favorable to innovation. The pursuit of this goal presupposes the existence of a solid commitment on the part of a large part of society, including not only the municipal authorities, but also the most representative segments and entities of society, the business class, NGOs and the academic community, in order to develop a digital city of mixed initiatives.

It should be borne in mind that the starting point in this field of technological convergence is the relationship between cities and the infinite possibilities opened up by technological advances and their applications in local development, boosting the economic, social, cultural and political production of the place inhabited by the citizen and complementing the role of the real city, with its cyberspace dimension. Network infrastructure is essential to absorb the demands of applying information and communications technologies.

The Ministry of Communications' Digital Cities Project for the state of Rio de Janeiro is still a long way from completion. Only the municipality of Engenheiro Paulo de Frontin shows signs of starting work at the end of March or April 2016.

In this sense, the presentation of other initiatives in the field of information and communication technology is necessary to identify the potential of these cities selected by PAC2 and the Pilot Project.

Digital cities are the expression of a networked society and imply preparing the territory to absorb the necessary infrastructure to interconnect the physical systems that relate to the citizen and integrate material and virtual space, transforming time into an indicator rather than a determinant.

Enabling widespread access to a network such as the internet requires initiatives in the field of digital inclusion that go beyond learning computers or enabling access. We need to create an impact strategy in the field of technological literacy. For this to happen, it is necessary to promote a relationship with local culture, so that information is no longer linear and unidirectional, but becomes interactive and enables effective participation in the field of citizenship and universal access to education.

5. Bibliographical references

AMARAL, Bruno do. **Digital cities project practically abandons the use of poles.** Available at: <http://convergecom.com.br/teletime/02/07/2015/projeto-de-cidades-digitais- praticamente-abandona-uso-de-postes/>. Accessed on: 27 Jan. 2016.

BORJA Y CASTELLS, Jordi Manuel, **Local Y Global, La Gestionde las ciudades en la era de la información**, Ed Taurus. 1997

BRAZIL. Constitution (2010). Executive Decree No. 7.175, of May 12, 2010. Establishes the National Broadband Program - PNBL; provides for the relocation of commission positions; amends Annex II to Decree No. 6.188, of August 17, 2007; amends and adds provisions to Decree No. 6.948, of August 25, 2009; and makes other provisions.... **Decree No. 7.175, of May 12, 2010**. Available at: <http://www.planalto.gov.br/ccivil_03/_Ato2007-2010/2010/Decreto/D7175.htm>. Accessed on: January 10, 2016.

BRAZIL. MINISTRY OF COMMUNICATIONS. (Ed.). **Ordinances.** various. Ordinances. Available at: <http://www.mc.gov.br/portarias?start=130>. Accessed on: March 12, 2016.

BRASILIA. Lygia Pupatto. **DIGITAL CITIES BUILDING AN ECOSYSTEM OF COOPERATION AND INNOVATION**. 2012. Available at: <https://www.google.com.br/url?sa=t&rct=j&q=&esrc=s&source=web&cd=1&cad=rja&uact=8 &ved=0ahUKEwiW4KDp8s_LAhXEWSYKHZ5GD1QQFggdMAA&url=http://www.mc.gov.br/d oc-crs/doc_download/1138-termo-de-referencia&usg=AFQjCNERFBFZYiwDNUe_54Eyq6bAG51d9g&sig2=6NMtSl2l4U-bnbclQKQAtw>. Accessed on: March 20, 2016.

BRASILIA. ANATEL. . **Licensing of Stations.** Available at: <http://www.anatel.gov.br/setorregulado/>. Accessed on: September 10, 2015.

Bom Jesus do Itabapoana**.** Available at: <http://www.bomjesus.rj.gov.br/site/>. Accessed on: March 10, 2016.

CASTELLS, Manuel. **The information age: economy, society and culture**. Sâo Paulo: Paz e Terra. 1999.

CASTELLS, Manuel. **o Poder da identidade.** Sâo Paulo: Paz e Terra. 1999.

CASIMIRO, de ABREU. **Data.** Available at: <http://www.casimirodeabreu.rj.gov.br/dados-municipais.html>. Accessed on: Feb. 10, 2016.

CIDADES.DIGITAIS@COMUNICACOES.GOV.BR, Digital Cities. **Information about the cities selected in the RJ digital cities program.** [personal message] Message received by: <micheletcs@gmail.com>. on: 25 Jan. 2016.

CPQD. **CPqD Technology Notebooks.** Available at: <http://www.cpqd.com.br/>. Accessed on: 03

Dec. 2015.

DE CINDIO, F., SCHULER, D.. Beyond Community Networks: From Local to Global, from Participation to Deliberation. **The Journal of Community informatics**, North America, 8, jun. 2012. Available at: <http://ci-ioumal.net/index.php/ciej/article/view/908/933>. Accessed: Feb. 12, 2016.

MINISTRY OF COMMUNICATIONS MINISTER'S OFFICE. Ordinance No. 4.699, of October 14, 2015. Amends Ordinance No. 376 of August 19, 2011, on the establishment of the Digital Cities Implementation and Maintenance Project. Brasilia, BRASILIA: Official Gazette, October 16, 2015. Section 1, p. 52-52. Available at: http://www.lex.com.br/legis 27031634 PORTARIA N 4699 DE 14 DE OUTUBRO DE 2 015.aspx . Accessed on: 10 Dec. 2015.

GUERREIRO, Evandro Prestes; SILVA, Michéle Tancman da. Digital Cities in Brazil - White Paper (preliminary version for debate). Sâo Paulo-SP, Brazil: SOCINFO, 2002.

HARVEY, David. Postmodern Condition. São Paulo: Loyola, 1993.

IBGE. **Population estimates for Brazilian municipalities and states.** 2015. Available at:

<http://www.ibge.gov.br/home/estatistica/populacao/estimativa2015/estimativa_dou.shtm>. Accessed on: March 12, 2016.

ITAOCARA. **Data.** Available at: <http://www.itaocara.rj.gov.br/>. Accessed on: Dec. 10, 2015.

JONAS GOMES DE CASTRO, **DIGITAL CITIES: AN ANALYSIS OF THE IMPLEMENTATION OF THE INFRASTRUCTURE IN ITS PILOT PROJECT, Brasilia**, Research report to be presented as the concluding work of the discipline "Residency in Public Policies". 2015

LORENZETO, Raphael de Abreu. **Maps.** Available at:

<https://pt.wikipedia.org/wiki/Wikipédia:Pàgina_principal>. Accessed on: 12 Dec. 2015.

Ministry of Communications Digital Inclusion Secretariat (Ed.). **Primer for Parliamentary Amendments to the 2016 Budget.** 2015. Available at:

<http://www.comunicacoes.gov.br/documentos/espaco-radiodifusor/cartilha-revisada.pdf>. Accessed on: 07 Jan. 2016.

Ministry of Communications. **Digital Cities:** Documents. Available at: <http://www.comunicacoes.gov.br/formularios-e-requerimentos/cat_view/22-acoes/27- cidades-digitais>. Accessed on: January 16, 2016.

Ministry of Communications (Ed.). **Parliamentary Amendments:** Digital Cities Project (Action: 00PA). 2015. Available at: <http://www.comunicacoes.gov.br/emendas- parlamentares>. Accessed on: Feb. 10, 2016.

Ministry of Communications (Org.). **Structure - Digital Inclusion Secretariat (SID).** 2014. Available at: <http://www.mc.gov.br/estrutura-institucional/142-institucional/secretaria-de- inclusao-

digital/32325-secretaria-de-inclusao-digital-sid>. Accessed on: January 14, 2016.

Ministry of Planning. **Digital Cities - Rio de Janeiro.** Available at: <http://www.pac.gov.br/infraestrutura-social-e-urbana/cidades-digitais/rj>. Accessed on: January 13, 2016.

PIRES, Hindenburgo Francisco. **Virtual structures of accumulation and cybercities. Scripta Nova - REVISTA ELECTRÓNICA DE GEOGRAFÌA Y CIENCIAS SOCIALES**, University of Barcelona. ISSN: 1138-9788. Legal Deposit: B. 21.741-98. Vol. VIII, no. 170 (59), August 1, 2004

SANTOS, Milton. Territory and Society - an interview with Milton Santos. Sâo Paulo: Fundaçâo Perseu Abramo, 2000, p. 60

SANTOS, Milton. **The Nature of Space. Technique and Time. Reason and emotion**. São Paulo. HUCITEC, 1996

SANTOS, Milton. **O recomeço da história,** Folha online, available at: http://www.uol.com.br/fol/brasil500/dc_3_1.htm. Accessed on 14/09/2014.

SEIDEL, E. Progress on "LTE Advanced" - the new 4G standard.2008. Available at: < http://www.nomor.de/root/downloads/white-paper/lteadvanced 2008 -07.pdf > . Accessed on: 12 Mar 2009.

Serpro. **Serpro cloud hosts applications for Digital Cities:** Service contracted by MiniCom will be used by 80 municipalities that are part of the program. Available at: <http://www.serpro.gov.br/noticias/nuvem-do-serpro-hospeda-aplicativos-para-o-cidades-digitais-1>. Accessed on: January 17, 2016.

SILVA, Michéle Tancman da, . **SCRIPTA NOVA REVISTA ELECTRÓNICA DE GEOGRAFÌA Y CIENCIAS SOCIALES: A (CYBER) GEOGRAFIA DAS CIDADES DIGITAIS.** Barcelona: Universidad de Barcelona, v. 2, n. 170, August 1, 2004. Annual. Available at: <http://www.ub.edu/geocrit/sn/sn-170-36.htm>. Accessed on: 01 Dec. 2015.

SYMANTEC. **Access point.** Available at: <https://www.symantec.com/pt/br/security_response/glossary/define.jsp?letter=a&word=acce ss-point>. Accessed on: Dec. 15, 2015.

SIMÔES, Helton Gomes; REIS, Thiago. **Broadband speed in Brazil varies between Libyan and Japanese rates:** Map shows disparity between cities; Internet turns 20 in Brazil. Average speed is 3 Mbps; government wants to increase to 25 Mbps by 2018. 2015. Available at: <http://g1.globo.com/tecnologia/noticia/2015/05/velocidade-da-banda-larga- no-brasil-varia-entre-taxas-de-libia-e-japao.html>. Accessed on: May 13, 2015.

TELEco: Wi-Fi and WiMAX ii: WiMAX concepts. Internet: Teleco, 16 Dec. 2015. Available at: <http://www.teleco.com.br/tutoriais/tutorialww2/pagina_2.asp>. Accessed on: 16 Dec. 2015.

MINICOM TV. **Digital cities project is one of the priorities of the Ministry of Communications.** 2012. Available at: <https://www.youtube.com/watch?v=7eV0Rtahx9o>. Accessed on: 08 Feb. 2016.

WOLLFENBUTTEL, andréa. **Technology in the vein - In Pirai, Rio de Janeiro, everyone has access to the internet.** 2005. IPEA. Available at: <http://www.ipea.gov.br/desafios/index.php?option=com_content&view=article&id=2182:clim a-responsabilidade-de-todos&catid=28&Itemid=23>. Accessed on: December 10, 2015.

Printed by Books on Demand GmbH, Norderstedt / Germany